Studying Scientific Me' ' in Translation

This book offers an interdisciplinary, multilingual, and data-rich investigation of how different types of metaphor are translated in scientific discourse. A treasure trove for scholars and students in both translation studies and metaphor studies which has long been overdue.

—Lettie Dorst, Leiden University, The Netherlands

Studying Scientific Metaphor in Translation presents a multilingual examination of the translation of metaphors. Mark Shuttleworth explores this facet of translation and develops a theoretically nuanced description of the procedures that translators have recourse to when translating metaphorical language. Drawing on a core corpus consisting of six *Scientific American* articles in the fields of neurobiology and biotechnology dating from 2004, along with their translations into Chinese, French, German, Italian, Polish and Russian, Shuttleworth provides a data-driven and theoretically informed picture of the processes that underpin metaphor translation. The book builds interdisciplinary bridges between translation scholars and metaphor researchers, proposes a new set of procedures for metaphor translation conceived within the context of descriptive translation studies and puts forward a possible resolution to the debate on metaphor translatability.

Mark Shuttleworth is Senior Lecturer at University College London.

Routledge Advances in Translation and Interpreting Studies

For a full list of titles in this series, please visit https://www.routledge.com

13 **Cultural Politics of Translation**
East Africa in a Global Context
Alamin M. Mazrui

14 **Bourdieu in Translation Studies**
The Socio-cultural Dynamics of Shakespeare Translation in Egypt
Sameh Hanna

15 **Ubiquitous Translation**
Piotr Blumczynski

16 **Translating Women**
Different Voices and New Horizons
Edited by Luise von Flotow and Farzaneh Farahzad

17 **Consecutive Notetaking and Interpreter Training**
Edited by Yasumasa Someya

18 **Queer in Translation**
Edited by B. J. Epstein and Robert Gillett

19 **Critical Translation Studies**
Douglas Robinson

20 **Feminist Translation Studies**
Local and Transnational Perspectives
Edited by Olga Castro and Emek Ergun

21 **Studying Scientific Metaphor in Translation**
An Inquiry into Cross-Lingual Translation Practices
Mark Shuttleworth

Studying Scientific Metaphor in Translation

An Inquiry into Cross-Lingual Translation Practices

Mark Shuttleworth

Routledge
Taylor & Francis Group
LONDON AND NEW YORK

First published 2017 by Routledge

2 Park Square, Milton Park, Abingdon, Oxon OX14 4RN
605 Third Avenue, New York, NY 10017

Routledge is an imprint of the Taylor & Francis Group, an informa business

First issued in paperback 2021

Library of Congress Cataloging-in-Publication Data
A catalog record for this book has been requested

ISBN: 978-1-138-93431-3 (hbk)
ISBN: 978-0-367-36632-2 (pbk)

Typeset in Sabon
by Apex CoVantage, LLC

To Tanya, Maria, Andrei and Alex, who have put up with so much because of this book.

Contents

Tables and Figure

Tables

Figure

Acknowledgements

My heartfelt thanks go to Charles Drage and Kirsten Malmkjær, who supervised this research before it metamorphosed from a PhD thesis into a book, and to Christina Schaeffner, Ian Press and Martyn Kingsbury, who were my examiners. I should also like to register my gratitude to the following people:

Charles Drage, once again, for his constant help, encouragement and erudition, and for helping me with the task of metaphor identification
Su Peneycad, for her help and support and for reading large portions of both versions of the work
Caiwen Wang and Tuan-Chi Hsieh for answering my questions relating to Chinese
Maggie Awadallah and Ilana Wartenberg for assisting me with the small amount of Arabic and Hebrew text that appears in the book
Maurice Mashaal from *Pour la Science* and Hartwig Hanser from *Spektrum der Wissenschaft* for the readiness with which they responded to my questions about editorial policy
The large number of people who provided support and advice, supplied copies of their theses or other materials, or helped me source *Scientific American* articles in the different languages included in the research
Needless to say, all my students, at the PhD, Masters and Bachelor levels, for their constant stimulation

And once again needless to say, any mistakes or deficiencies are my own responsibility.

I should like to acknowledge with thanks the following permissions:

Taylor and Francis, for allowing me to reproduce extracts from 'Translational Behaviour at the Frontiers of Scientific Knowledge' that appeared in *The Translator*, 17:2, 301–323 (Shuttleworth 2011).

Bononia University Press, for permission to reproduce parts of 'Translation Studies and Metaphor Studies: Possible Paths of Interaction between Two Well-Established Disciplines', in Donna R. Miller & Enrico Monti (eds), *Tradurre Figure / Translating Figurative Language*, Quaderni del Centro di Studi Linguistico-Culturali, Atti di Convegni CeSLiC—3, Bologna: AMSActa (available at http://amsacta.unibo.it/4030/), 53–65 (Shuttleworth 2014).

Lucile Desblache, the editor of the *Journal of Specialised Translation*, for allowing me to include an amended version of my article 'Scientific Rich Images in Translation: A Multilingual Study', in *JoSTrans, Journal of Specialised Translation* (2014) 21: 35–51.

Mary Snell-Hornby, for permission to reuse the image that appears on page 59 of the 1995 edition of her *Translation Studies: An Integrated Approach* (Figure 3.1).

Note

As discussed later on, in this book I am following the convention of marking all mappings in small capitals and verbalising them in the form of A IS B statements. I am also using bold to indicate the metaphorical elements within an example and underlining to show non-metaphorical renderings of source-text metaphorical expressions.

Introduction

Setting the Scene

If you travel on the roads of Greece, you cannot fail to notice the large number of lorries bearing the word *Metaforá* or *Metaforikí* on their sides. This is by no means an indication that Greek hauliers have a corporate interest in figurative language (however appealing that idea might be), but simply tells you that the vehicle is involved in *transporting* goods between locations. Along similar lines, the other main keyword in this investigation, *translation* (together with a number of other Latinate alternatives that were rejected along the way: see Pym 2000:108–31), clearly conveys this same meaning of *movement* from one place to another. Thus anyone foolhardy enough to write about metaphor in translation finds him or herself obliged to think simultaneously in terms of two separate types of meaning transfer.

The origin of the word *metaphor* is indeed Greek (*meta-* 'change (of place, order, condition, or nature)' and *pherein* 'to bear/carry'), while *translate* comes from Latin (*translātus*, the past passive participle of *transferre* 'to transfer', from *trans-* 'across, to or on the farther side of, beyond, over' and *ferre* 'to bear/carry') (*The Shorter Oxford English Dictionary*). The idea of transfer is also reflected in the terminology of both metaphor research and translation studies, as evidenced in the twofold use of the terms *source* and *target*. In the former discipline, these two terms respectively refer—in one major theory of metaphor at any rate—to the two domains of experience brought together in a metaphor, the one providing the words and concepts in terms of which the other is spoken, written or (possibly) even thought about, while in the latter discipline they denote the original and receptor texts, languages or cultures between which the act of translation takes place.

So etymologically and conceptually, the notions of metaphor, transfer and translation are closely related.

Along similar lines, the concept of literalness also takes on a double meaning: privileging source-text wording over sense in relation to translation and non-figurative over figurative means of expression in relation to metaphorical language. 'Literal' language is a goal that science has sometimes consciously striven after in the belief that a metaphorless discourse should be preferred in order to be sure of conveying scientific knowledge in a manner that is totally neutral. In line with much modern research into metaphor, however, it is the contention of this work that metaphorical modes of expression and terminology formation are the norm rather than the exception in virtually all forms of human communication, including the language of science. But surely there is no metaphor in science? Surely there is a great gulf set between the discourse of hard science and that of other kinds of discourse that are not intended to be the conveyors of impartial truth? Let us be under no illusion that all scientific prose is by necessity dry and functional: the texts that form the corpus for this research—which could be characterised as belonging to the difficult end of popular science—display a conglomeration of literary, often playful, textual features including quotation, allusion, new word coining, word play and, of course, metaphor. Indeed, it is a contention of this research—and this is something that emerges from the data—that metaphor appears to be an essential part of the processes of science communication and scientific concept formation.

Given the potential of metaphor for alternative, often fluid, interpretation, it is an interesting irony of modern metaphor theory that some of it is surprisingly rigid. Some of the best-known writers within the approach known as the conceptual metaphor theory have tended to be very dismissive of other approaches, grouping them all together under the disparaging title of 'traditional'. While the approach to metaphor taken in this research is broadly in alignment with this theory, it employs a version of its theoretical framework that is more suited to the study of translation.

About the Research

One of the original reasons for my interest in metaphor in translation was the Hungarian composer György Ligeti—a monograph on whose music I translated during the late 1990s (Lobanova 2002; see also Shuttleworth 2013). What struck me very forcefully about this text was the way in which the whole discourse was shaped by a number of unusual underlying metaphors. Reflecting on the translation process both while it was in progress and after it had been completed led me to conclude that the reason for the considerable difficulty involved in translating the text was its deep-rooted use of metaphor. Since that time, the subject has held a special interest for me.

The research on which this book is based involves the study of the many and varied types of metaphorical expression contained in scientific texts and their translations. The research aims to employ a range of concepts from metaphor research to examine what happens to metaphor in translation, and it is believed to be one of the first major studies of metaphor in translation that is centred on modern scientific discourse. It is based on a corpus of six articles that appeared in *Scientific American* during the first half of 2004—and also in French *Pour la Science* ('For Science'), Italian

Le Scienze ('The Sciences'), German *Spektrum der Wissenschaft* ('Spectrum of Science'), Polish *Świat nauki* ('The World of Science'), Russian *В мире науки* (*V mire nauki*, 'In the World of Science') and Taiwanese Traditional Chinese 科學人 (*Kexueren*, 'The Scientist') over the ensuing months—which was around the time when this research originally got underway in its present form. I selected this journal largely because it is translated month by month into 13 languages, including the 6 languages that are being focused on in this research—Chinese, French, German, Italian, Polish and Russian. Over the years, *Scientific American* articles have been studied by a number of translation scholars (see, for example, Rey 2000; Bowker & Pearson 2002; Hoorickx-Raucq 2005; Liao 2007, 2010, 2011; Sharkas 2009; Chan 2011; Liu 2011) and offer a rich resource in terms of the quantity of material and the number of target languages that are included. The research is, however, in a sense incomplete as not all the different language editions have been studied.

The addition of a further target language (Chinese) was one of the major changes that this work underwent as part of the process of converting it from a doctoral thesis to a monograph. This proved extremely fruitful, as it forced me to think afresh about questions of analysis, categorisation and presentation, hopefully from a more mature standpoint than before.

Broadly speaking, the book aims to investigate translation phenomena, on a multilingual basis and relating to popular science texts within the broad area that I refer to as 'metaphor in translation', and to probe the interrelations that exist (or can be established) between translation studies and metaphor research. The considerable number of data categories that the research has generated has meant that not everything can be illustrated in detail—or indeed at all, in some cases. Throughout, the analysis depends on the idea that a particular translation solution can simultaneously be seen in multiple different contexts and therefore as representing a variety of context-specific procedures.

In the work, a number of branches of life science are being taken as representative of scientific discourse in general. So on what basis can the book claim to be generalising about 'scientific metaphor', as implied in the title? It has to be said that one of the reasons for this has to do with the research methods used. Focusing on specific areas allows for greater coherence in the analysis in terms of the specific metaphorical expressions that are identified and helps to ensure that larger metaphorical structures are represented by a sufficient number of examples. Be that as it may, if you were to carry out a study of the linguistic characteristics of the discourse of bioscience, then you would likely identify some features that were common to the general language of science and others that were more specific to this particular subject area, with this pattern being mirrored on the metaphorical plane.

A related point concerns the choice of *Scientific American* as the source of data for the project. Given that it is not a peer-reviewed journal and is intended for a non-specialist readership, and it is thus not a forum in which original scientific advances first see the light of day, can we be certain that the material studied relates to the technical cutting edge of science?

Once again, there is a simple, practical reason behind the choice of *Scientific American* as the source of data for this project, which is its profile as a well-known, prestigious science journal that is published in many international editions. That aside, though, in spite of its status as a journal of popular science, its authors count well over a hundred Nobel Prize scientists among their number. So what are the credentials of a typical *Scientific American* article as a vehicle for the very latest scientific thought? In most instances, articles are written by scientists who are presenting their own recent research. In the case of the six articles forming the basis of this study (which are listed later on in the introduction), five are written by the scientists themselves and only one ('Synthetic Life') by a 'senior writer'. Each of the six either describes research that is currently being conducted by the authors themselves, or else references or acknowledges recent research publications, once again frequently written by the article authors. Thus, for example, 'The Addicted Brain', which appeared in March 2004, acknowledges work published at the beginning of 2003 in *Neuroscience*, and 'Synthetic Life', published in the May 2004 issue, references a *Science* article from August 2003. In brief, the topics of the articles are as follows:

- 'Decoding Schizophrenia' (Javitt & Coyle 2004): how an improved understanding of signalling in the brains of people suffering from this disorder promises the development of better therapy
- 'The Addicted Brain' (Nestler & Malenka 2004): how understanding the long-term effects of addiction on the brain could improve treatment of the underlying compulsive behaviours
- 'The Other Half of the Brain' (Fields 2004): the newly understood importance of glial cells—a very common type of cell in the central nervous system—for thinking and learning
- 'Evolution Encoded' (Freeland & Hurst 2004): how nature's 'programming' helps prevent catastrophic errors occurring through the rules governing how genes encode proteins
- 'Synthetic Life' (Gibbs 2004): the creation of programmable, living 'machines' out of DNA parts assembled inside microbes
- 'Freud Returns' (Solms 2004): how Freud's theories are being found to provide a good explanation for neuroscientists' biological descriptions of the brain

In other words, the data on which the research is based relates to the broad subject areas of neurobiology, biotechnology, genetics and psychology; taken as a whole, it offers significant insight into the metaphorical patterns associated with nature, genetic material and the 'wiring' of the brain, if only because of the large number of available examples that relate to these areas.

The research appears to bear witness to a subtle interplay between the implementation of conscious translation procedures by translators and the powerful gravitational pull of source-language patterns that may give rise to an unwillingness to seek out functional equivalents in the target language. This is mirrored by another type of interplay between the predictable and the one-off. Both these different kinds of interplay serve to characterise translators' approach to metaphor translation and create the space within which the generalisations put forward later on in the book can emerge.

The work's multilingual nature is intended to bring to the fore what is typical for translator behaviour across a range of different linguistic and cultural contexts, and it views the data from a range of different angles, employing the broad framework of descriptive translation studies as well as different concepts from metaphor research. Because of this emphasis, I have not simply devoted a chapter or section to each target language, but instead have tried as a rule to consider examples from multiple languages together in order to see what patterns emerge. However, the book is ultimately about translation—not a transparent glass or non-distorting mirror, nor an imperfect process that fails to convey an immutable 'meaning', but the real-world attempt on the part of translators to reformulate complex significations using the linguistic resources of the target language along with their own reserves of ingenuity.

The characteristics of translated as opposed to original text are of great importance in the research, as is what happens specifically to metaphors and, more importantly, metaphorical expressions in translation. My work is based on the conviction that metaphor pervades all kinds of discourse and that it plays a significant role in concept formation and even in the progress of scientific ideas and understanding. In terms of translation issues, I am interested in the effect that translators' micro-level decisions have on the overall configuration of metaphors in a text, and the specific procedures typically adopted by translators when rendering source-text metaphors and metaphorical expressions into a target language.

Because of the only patchy availability of data on the number and identity of the translators working in each language, and also in view of the likelihood of there being a relatively large number of them in most cases, the book is also not about individual translators and the choices that characterise their translating style. In addition, the context in which the translation takes place in the international editions of *Scientific American* is quite clearly an institutional one. The (for the most part anonymous) translators do not work in isolation and are presumably all constrained by local editorial policy and, possibly, by active editorial intervention.

Some Problems and Observations

Metaphor, it can be argued, is a particularly elusive feature of language. Experts disagree about what does or does not constitute a metaphor, and there is certainly no dictionary of all the metaphors that a particular language system can be said to 'contain'. While many metaphorical expressions—such as those that are referred to as 'conventionalised' later on in the book—may find a place in such a volume if it were ever produced, it could never be considered complete, as new metaphorical expressions are constantly being dreamt up on the spur of the moment to fit the requirements of a one-off communication context. With this in mind, what I am searching for in this research is expressions on the surface of the discourse that reflect particular metaphorical 'mappings', whether or not these latter account for a large number of expressions or just a single one.

The kind of inductive language-based research conducted in this study inevitably contains elements of subjectivity. First of all, messy, real-world candidate expressions need to be either accepted or rejected as metaphorical by the exercising of judgement. Secondly, the correct verbalisation for the mappings that are identified has to be worked out on the basis of imprecise criteria. Thirdly, as will be discussed in the following section, careful thought is required to find the most appropriate form for the numerous back-translations of examples that the book contains. Finally, multiple judgements need to be made regarding the categorisation of each original and translated example. Needless to say, in each of these areas, there is some room for disagreement.

Inevitably, you occasionally have to use metaphor to talk about metaphor. For example, if we use the text of this book as a corpus, it appears that the following metaphorical items all collocate with the word *metaphor*: the human qualities of *strong*, *bold*, *creative*, *guiding*, *live* and *dead*; the significant positional adjective *underlying* that suggests the kind of relationship that exists between conceptual metaphors and metaphorical expressions; and two modifiers, *extended* and *network*, that relate to how metaphors exist within texts. Such metaphors are of course difficult to avoid entirely; as stated by Mooij, 'It is a matter of course that explications of metaphor often show a certain groping for words and that in such explications new metaphorical expressions are often inevitable' (1976:104), even though there is probably nothing new in the metaphorical expressions listed earlier.

Along similar lines, it is also possible to observe a tendency to use humanising expressions when talking about specific languages or about language in general. Languages *permit*, *tolerate* or even *sanction* metaphor, for example. While I have tried not to use metaphorical expressions excessively, it would be by no means easy to avoid them completely. Apart from anything else, as I hope to demonstrate in Section 1.2.2, metaphors are a potential stimulus to new research developments and knowledge creation, so cultivating a completely metaphor-free discourse may ultimately prove to be a somewhat sterile approach.

A Note on Back-Translation

Following standard practice, the book makes use of back-translation as a procedure for illustrating the meaning of text in a language that is very possibly unknown to the reader. To my mind, its use is always potentially problematic, not least because it entails producing a new instance of the very phenomenon that one is supposed to be investigating—in this book, a translated metaphor. Most but perhaps not all back-translations will involve retaining the metaphorical expression, possibly in a slightly modified form. However, as far as this is possible, extreme care should always be taken to ensure that no new metaphors are introduced during the back-translation process, as these would create something of a distraction. All back-translations included in the book (bar two from Arabic) are my own.

Chapter Outlines

This introduction has been devoted to setting the scene, providing some preparatory discussion and making a number of further remarks of an introductory nature. It is followed by seven chapters that introduce the theoretical framework, present a detailed analysis of the data and provide some conclusions to the research.

Chapter One, 'Metaphor in Scientific Thought and Writing', discusses the place of metaphor in science and why it forms an essential part of it. Specific topics that are focused on include the role that metaphor plays in stimulating the development of scientific thinking and in forming scientific terminology. This leads to Chapter Two, 'Translating *Scientific American*', in which we take a look at the journal from which my data is drawn: a high profile popular science periodical that appears every month in multiple international editions. Each issue has a complex structure, and the translation policy used is seen to differ from edition to edition. The introductory part is brought to a conclusion in Chapter Three, 'Metaphor and Translation'. The first section of this chapter provides an introduction to the theoretical study of metaphor, focusing largely on the conceptual metaphor theory. In the second section, the discipline of translation studies is introduced, after which a selection of the work that has been done on metaphor in translation within the discipline is presented. The chapter is largely theme based and provides a historical overview of many of the topics discussed. Finally, section three considers a number of practical methodological questions and presents the multilingual approach that is being used in the book.

The next three chapters, and the two interludes, are taken up with a detailed discussion of scientific metaphor in translation. This analysis aims to track metaphor in translation on both macro- and micro-levels, and uses each of six theoretical 'parameters' (existence category, mapping, purpose, conventionality, provenance and richness) in turn to analyse what happens to the metaphorical language contained in the source texts when it is translated into the six target languages included in the project. A pair of related parameters are presented in each chapter, while the two short topic-based interludes are included to introduce a subsidiary focus organised along thematic rather than theoretical lines. Throughout these chapters, language-specific matters are discussed in passing as the need arises.

Multiple categorisations of procedures are offered, although these are not always intended to be exhaustive. While they are structured largely in parallel, the chapter sections are not all carbon copies of each other, as some of them pursue aims that are specific to one or other of the parameters that they are considering. Finally, for reasons of space, it has not been possible to supply an example for every single phenomenon that is discussed. Some examples are brief and others more extended, depending on their interest and the range of issues that they raise. Specific examples are sometimes repeated in different contexts throughout the book, as they may provide good illustration of more than one particular phenomenon.

Chapter Four, 'Macro-Level Metaphors', focuses on the so-called 'existence categories' (concretisation, humanisation, etc.) and the notion of mappings, both of which offer sets of high-level semantic categories for classifying metaphor. The first of these is relatively informal and is ultimately derived from the Great Chain of Being, while the second represents a central concept in the conceptual metaphor theory. This chapter is followed by Interlude One, 'Metaphors of Nature', in which I look at the wide range of metaphorisations of nature that are present in my data. Chapter Five, 'Intuitive Classifications of Metaphor', takes as its two standpoints the different scientific and other purposes that metaphorical expressions serve in the texts, and the concept of conventionality, which relates to the manner and extent to which a particular metaphorical expression is embedded in the language in which it occurs. The second chapter section also includes a discussion of Kloepfer's boldness hypothesis and a tentative conclusion as to its accuracy within the context of my research. Interlude Two, 'Metaphors of Genetics', then paints a picture of the multiple metaphors that are used in the source and target texts in order to talk about this area of scientific study. Chapter Six, 'Lakoff and Johnson's Metaphor Types', which forms the last part of the analysis contained in the book, presents the 'provenance' and 'richness' parameters. Provenance—which concerns whether a metaphorical expression is substantially image-schematic in nature, is based on propositional knowledge, or is an image metaphor—forms the subject of both chapter sections, with richness—which indicates how detailed and 'rich' in associations a metaphorical expression is—being presented as part of the first. This chapter includes what is probably the most comprehensive general list of procedures that the book contains and revisits Al-Harrasi's list of procedures for translating metaphor, which is discussed in Section 3.2.2.5.

Finally, Chapter Seven, 'Conclusion', presents the study's major findings, which include some proposals regarding how translation studies and metaphor research might interact, a summing up of the changes that metaphorical language typically undergoes in translation and a list of procedures for metaphor translation. The book ends with some suggestions for further research.

1 Metaphor in Scientific Thought and Writing

> Nonscientists tend to think that science works by deduction, [. . .] but actually science works mainly by metaphor.
>
> (Brian Arthur, quoted in Waldrop 1994:327)

> [. . .] more metaphorical analyses should be conducted on actual scientific texts.
>
> (Johnson-Sheehan 1998:177)

1.1 A Brief Introduction to Scientific Metaphor

It is by no means universally accepted that metaphor and science are inextricably linked. Many simply incorrectly believe that scientific discourse contains no metaphors—which in a sense is hardly surprising, given what we hear about the scientific method and science's commitment to transparent objectivity. More significantly, perhaps, there are some voices, both past and present, that may have acknowledged the existence of metaphor in scientific discourse, but have nonetheless advocated a careful avoidance of all metaphorical modes of expression. These positions will be considered briefly before we go on to discuss less inimical views on the function of metaphor in communicating scientific ideas. Finally, and building on the striking claim of the first epigraph to this chapter that metaphor does not simply exist in scientific texts but is the main mechanism that enables science to work at all, this section will consider the strong likelihood that metaphor serves the vital function of channelling scientific thought in particular directions and thus exercises a definite influence on the ways in which progress occurs.

1.1.1 Attempts to Dispense with Metaphor

As Ortony observes, science is 'supposed to be characterized by precision and the absence of ambiguity, and the language of science is assumed to be correspondingly precise and unambiguous—in short, literal' (1993b:1). Scientific texts are thought by some to be characterised by the use of clear, precise, unambiguous language suitable for reflecting the unembellished truth of science, with metaphors occurring rarely, if at all.

This notion at least partly takes its roots from the seventeenth century. The rise of science during this period coincided with a great interest on the part of many philosophers in the concept of an 'ideal' or 'perfect' language, which was conceived as a tool for representing concepts in an idealised manner without the interference of an intervening natural language. It was in 1678 that Leibniz produced a long fragment in which he proposed his *Lingua generalis*, one of a number of projects that emerged in the seventeenth century whose aim was the creation of an artificial linguistic apparatus, designed on purely logical terms, that would make this kind of direct representation of reality possible (Eco 1995:269–88).

While these experiments were abandoned as impracticable, the thought that they engendered, that a precise representation of the real world through the use of language but unobscured by ambiguity, fuzziness and figurative expression, has persisted in some quarters. The requirement for scientific texts to be characterised by the use of precise and unambiguous language was a distinctive quality of logical positivism, for example, which held that reality could only be exactly described through the medium of language 'in a manner that was clear, unambiguous, and, in principle, testable' (Ortony 1993b:1). The opposing view to this contends that the objective world 'is not directly accessible but is constructed on the basis of the constraining influences of human knowledge and language' (1993b:2). Ortony distinguishes two separate approaches to metaphor in line with these different ways of understanding how scientific language should function. The logical positivist view described earlier would be associated with an understanding of metaphor as something that is 'deviant and parasitic upon normal usage', and characteristic of rhetoric rather than science, while the contrary standpoint would be that metaphor was 'an essential characteristic of the creativity of language' (1993b:2; see also Leane 2007:83–4). In each case, my research aligns itself with the latter view—a position that is in line with modern research into language, cognition and the ability of human beings to process information.

1.2 Metaphor in Science Today

The last few hundred years have seen an abundance of scientific metaphors—including the medieval concept of the Book of Nature, Darwin's natural selection and the curved space of Einsteinian physics—and this is a tendency that continues until this day, in spite of the efforts of the logical positivists and others. This section provides an overview of metaphorical use in scientific discourse, focusing in particular on the question of how metaphorical language can perhaps influence the direction taken by scientific researchers and on how metaphor lies behind much scientific terminology.

1.2.1 Metaphor in Scientific Discourse

Interestingly, nowadays, the important role of figurative language and thinking in science is for the most part even acknowledged by writers from whom a highly cautious attitude to such matters would be expected:

> Metaphors and analogies are essential to science and theory. Complex and more abstract areas of science rely particularly on metaphor and analogy to add clarity to knowledge and to communicate that knowledge. This is perfectly legitimate and indeed, to some extent, unavoidable. In science, analogies and metaphors may emerge as useful ways to think about, describe, and explain objective facts and evidence. For example, psychologists have employed the metaphor of visual selective attention being like a 'spotlight' illuminating the relevant information out there in the world from the surrounding darkness of all that we ignore. In many respects this has proved a very fruitful metaphor guiding thinking in this area of study. The problem here is not the use of analogies or metaphor in scientific thinking, but the clear abuse of them.
>
> The problem with pseudoscience is its use and over-reliance on metaphor as an argument in and of itself. Rather than employ metaphors and analogies as illustrations of scientific knowledge, pseudoscience employs analogies to deduce new conclusions and propose alternative truths. At this point it no longer becomes a mere illustration; it becomes an argument by analogy (or metaphor . . .).
>
> (Braithwaite 2006)

It should be noted that Braithwaite argues that metaphor and analogy are sometimes needed in order to add clarity. Of interest here too is the distinction that he draws between science and pseudoscience in terms of their use of metaphor and analogy. The fact that this article was originally published by an organisation called UK-Skeptics means that the appraisal that it offers is likely to be a relatively sober one.

Mithen similarly identifies the use of metaphor and analogy as one of three critical properties of science (1996:245). However, it seems highly probable that the level of metaphoricity varies from one area of science to another. Dunbar, for example, argues that metaphors occur most frequently in texts about physics and evolutionary biology, with the reason being that the subject matter of these disciplines concerns phenomena 'that everyday experience does not equip us to talk about', unlike that of chemistry or anatomy, for example, for which the 'conventional mechanistic terminology' of everyday language is totally appropriate (1995:142). Dunbar also argues that such metaphors tend to use the social human world as their source domain (1995:142; see also Mithen 1996:308). Some brief comments will be made in Section 1.2.4 regarding how popular science texts differ from specialist ones in this respect.

Finally, it should be pointed out that some metaphors become very central to our way of thinking and speaking about certain subjects, which means that, if the time ever arrives when these metaphors come to be considered obsolete, a certain realignment of concepts and means of expression will become necessary. This is the case with evolutionary biology, very central to which has always been a metaphor originally suggested by Darwin himself in *The Origin of Species*: 'The affinities of all the beings of the same class have sometimes been represented by a great tree. I believe this simile largely speaks the truth' (Darwin 1872:129). Darwin's original vision was of the totality of life on earth represented by a tree, each branch of which was a single species, some of which reach an eventual dead end but others—today's surviving species—extend until the very top. This image has supplied several generations of scientists with not only a fundamental 'unifying principle for understanding the history of life on Earth' (Lawton 2009:34) but also with an ultimate aim in the form of the eventual faithful reconstruction of the tree itself.

However, this view has been gradually dismantled as we have improved our ability to read genetic material and, ultimately, entire genomes. In 1999, Doolittle made the provocative claim that 'the history of life cannot properly be represented as a tree' (1999:2124). Clearly, the Tree of Life does not exist in nature, but it is rather imposed on nature as a framework for classification (see Lawton 2009:37). If species do not simply pass on traits but also regularly exchange genetic material or hybridise with other species (as it appears that they do), then what emerges is not a 'neat branching pattern' so much as an 'impenetrable thicket of interrelatedness' (2009:36). While the model that it represents has probably not yet outlived its usefulness when applied to animals and plants, in all likelihood it no longer provides an adequate description of the workings of evolution in general; like Newton's mechanics, it has proved to be revolutionary and highly fruitful in its time but probably can no longer account for the highly complex data that is now being observed and discovered in the real world (see Lawton 2009:38–9).

In addition, notwithstanding these assertions by writers such as Doolittle and Lawton, the concept of the Tree of Life is still very much alive in microbiology and molecular biology, although as a pictorial representation it is undergoing constant development and revision. Modern versions of the Tree of Life in fact no longer really resemble a tree, and the original terms 'tree' and 'branch' have now been joined by others such as 'lineage' (see Pace 2009; Hug et al. 2016).

Interestingly, metaphorical language can sometimes start to fall behind current scientific thought, eventually even lagging so dramatically that it can actually belie the scientific worldview, as we shall see in Interlude One. However, the next section will consider the implications of this relationship between science and metaphor in somewhat more general terms.

1.2.2 *Metaphor and the Channelling of Scientific Thought*

What Braithwaite does not appear to allow for is the possibility that metaphor's scientific scope of application is not limited to conceptualising, describing and explaining objective facts and evidence but that it can play a role in helping our understanding of the natural world to advance (selectively) in certain directions. In the technical literature, this is known as metaphor's theory-constitutive function, and it will form part of our discussion later on (see Section 5.1.1).

We have already considered the example of how Darwin's tree metaphor has acted as a convenient framework for conceptualising the workings of evolution and how it gradually appears to be giving way to alternative models—or at least transforming itself beyond recognition—as our understanding of the relevant processes has increased. As I have just suggested, while it is current, such a metaphor will suggest to researchers certain avenues of investigation that are congruent with it and at the same time may make it less likely that other directions will be pursued. As another possible instance of this phenomenon, we may cite the metaphorical use of the word *engineering* to refer to technical areas to which it did not previously apply—for example, energy engineering, nanoengineering, genetic engineering and language engineering. It seems that this *engineering* metaphor, once adopted, will influence the way scientists conceptualise their work and may very likely lead their research in new directions.

If such a channelling influence does indeed exist, we would expect scientists themselves to articulate their thoughts on it to some extent. Interestingly, the author of at least one of the articles included in the corpus does indeed reflect on this phenomenon. The following quotation, for example, comes from the article entitled 'Synthetic Life':

> 'Interchangeable components are something we take for granted in other kinds of engineering,' Endy notes, but genetic engineering is only beginning to draw on the power of the concept. One advantage it offers is abstraction. Just as electrical engineers need not know what is inside a capacitor before they use it in a circuit, biological engineers would like to be able to use a genetic toggle switch while remaining blissfully ignorant of the binding coefficients and biochemical makeup of the promoters, repressors, activators, inducers and other genetic elements that make the switch work. (Gibbs 2004:77)

There are a number of further examples from Gibbs (2004) that also imply an acceptance of the legitimacy of this function of metaphor, including the following:

> Biologists are crafting libraries of interchangeable DNA parts and assembling them inside microbes to create programmable, living machines. (Gibbs 2004:75)

> 'We would like to be able to routinely assemble systems from pieces that are well described and well behaved,' Endy remarks.
>
> (Gibbs 2004:76)

> That way, if in the future someone asks me to make an organism that, say, counts to 3,000 and then turns left, I can grab the parts I need off the shelf, hook them together and predict how they will perform.
>
> (Gibbs 2004:76)

> At a mechanical level, individual BioBricks (as the M.I.T. group calls the parts) can be fabricated and stored separately, then later stitched together to form larger bits of DNA.
>
> (Gibbs 2004:77)

We can also find such an acceptance in other *Scientific American* articles, such as the following:

> . . . once nature developed such an effective mechanism for silencing the subversive genes in viruses and mobile DNA sequences, it started borrowing tools from the RNAi tool chest and using them for different purposes.
>
> (Lau & Bartel 2003:40)

Taken together, these examples represent strong evidence for the validity of the idea that the existence of a metaphor can in some way engender particular kinds of creative thinking in the practitioners of a specific scientific discipline. However, as conceded by Gibbs, such analogies can only be applied in certain respects (2004:78)—an observation that is of course in line with the concept of partial mapping that is introduced in Section 3.1.2.

1.2.3 Metaphor and Terminology

From at least the time when the eleventh-century monk Constantine the African, working on metaphorical principles, used the Latin word *cataracta* ('large waterfall') to describe a rather nasty condition of the eye (McVaugh 2001:326–8), metaphor has been used as a basis for scientific and medical terminology. It is in fact one of the main ways in which new terms are formed (besides morphological, syntactic, lexical and stylistic: see Divasson & León 2006:59, 61). As will be discussed later in this section, ancient languages such as Greek and Latin play an important role in terminology formation in many Western languages, which means that in any terms created in this manner, the metaphoricity will be hidden, although still implicitly present.

As argued by Arbib, 'Almost any interesting descriptive term can be shown etymologically to be a dead metaphor', the examples that he provides being *spirits* (for whisky), *leaves* (for pages in a book) and *fiery* (for a person's temperament) (2012:277). A similar observation had in fact been made by Owen Barfield—a member of the famous Oxford-based Inklings literary group—more than 80 years earlier:

> . . . one of the first things that a student of etymology . . . discovers for himself is that every modern language . . . is *apparently* nothing, from beginning to end, but an unconscionable tissue of dead, or petrified metaphors. . . . If we trace the meanings of a great many words . . . about as far back as etymology can take us, we are at once made to realize that an overwhelming proportion, if not all, of them referred in earlier days to one of these two solid things—a solid, sensible object, or some animal (probably human) activity.
>
> (1973:63–4, first published 1928)

This observation quite clearly goes much further than merely commenting on the derivation of scientific metaphor. However, the extent to which a term's (or a word's) etymology—whether it is transparent or opaque—will affect a reader's interpretation of it (or his or her ability to identify it as essentially metaphorical) will be considered briefly at the end of this section. The prevalence of dead metaphor can create something of a dilemma for the translation researcher. How much of this should be included in the data: all, some or none of it? If every example is included then it is possible that other, more 'interesting' types of metaphors may be drowned out, and yet without it the picture will be incomplete. In my research, I have tried to resolve this issue by including a representative sample of examples, some of whose metaphorical import is contained in their Greek or Latin roots, but leaving dead metaphor largely to one side, working on the assumption that its translation is probably not of interest to translation studies, if only because in most cases it tends to be largely automatic.

Many Western European languages in fact contain considerable amounts of terminology derived from these ancient tongues. Indeed, according to Casselman, such terms account for a staggering 98% of the English scientific vocabulary (1998:vii). The main reason for this is the millennia-long tradition that has existed in the West that has treated Latin—and to a lesser extent Greek—as the only proper media for transmitting knowledge and learning (1998), but it is also possibly due in part to the tendency to suppress metaphor that was referred to in Section 1.1: while not removing a metaphor entirely, this approach to terminology formation will at least conceal a term's metaphoricity within the Latin or Greek morphemes of which it is composed.

Finally in this section, we need to turn to the question of the relative transparency of scientific terms in different languages. The morpheme-by-morpheme meaning of terms can either be largely concealed, as is the case with Latin and Greek terms in English (assuming that the reader has no knowledge of these languages), or partially or completely apparent, as with many terms in German, Polish, Russian and Chinese.

In order to illustrate this, we will look at three brief examples. The first two, which are taken from Hofstadter (1997), focus exclusively on Chinese, while the third, which I have produced myself, is multilingual.

The first example concerns the Chinese words for different kinds of dinosaur. As is well known, in common with most or all other European languages, English derives its words for these ancient creatures from Greek roots—the precise meanings of which are probably not available to most speakers unless they take the trouble to find them out. In contrast, the meanings of the Chinese terms are often completely transparent to native speakers, as particular characters with the required ranges of meaning are simply collocated to produce a new word. In many but not all cases, the meanings are direct translations of the Greek terms used in Western languages (see Hofstadter 1997:297; the simplified Chinese form is given first, followed by the traditional):

brontosaurus ('thunder lizard'):	雷龙/雷龍 (*leilong*, 'thunder dragon')
tyrannosaurus ('tyrant lizard'):	霸王龙/霸王龍 (*bawanglong*, 'tyrant-king dragon')
pterodactyl ('wing finger'):	翼手龙/翼手龍 (*yishoulong*, 'wing-hand dragon')
stegosaurus ('roof lizard'):	剑龙/劍龍 (*jianlong*, 'sword dragon')
triceratops ('three-horned face'):	三角龙/三角龍 (*sanjiaolong*, 'three-horn dragon')

We see precisely the same situation with these names for subatomic particles (1997:298):

electron ('amber' + suffix -on):	电子/電子 (*dianzi*, 'spark seed')
proton ('first' + suffix -on):	质子/質子 (*zhizi*, 'primal seed')
neutron ('neutral' + suffix -on):	中子/中子 (*zhongzi*, 'neutral seed')
neutrino ('neutral' + suffix -in + suffix -o):	中微子/中微子 (*zhongweizi*, 'neutral seedlet')
photon ('light' + suffix -on):	光子/光子 (*guangzi*, 'light seed')

In this way, in this instance at least, Chinese probably represents the extreme of transparency, while in a language such as English, we see something approaching a maximum level of opacity.

Finally, Table 1.1 presents a multilingual comparison of the names of gases, including not only English and Chinese but also some of the other languages focused on in my research. The meanings of the individual morphemes of the German and Russian terms parallel those of the English, with the important difference, of course, that they are quite transparent because they are based on native rather than classical etymologies. The Polish, interestingly, uses a variety of approaches: it creates new words for oxygen and hydrogen from its own word stock but borrows the word for nitrogen from French (which itself is

Table 1.1 Relative Transparency of Terms for Common Gases in Different Languages (as before, the simplified Chinese form precedes the traditional form)

English	*Polish*	*German*	*Russian*	*Chinese*
oxygen (from Greek *sharp/acidic* + *brought forth*)	tlen (from *tlić się* = to smoulder)	Sauerstoff (*sour* + *substance*; partially calqued on the Greek)	кислород (*kislorod*; *acidic* + *origin*; calqued on the Greek)	氧气/氧氣 (*yangqi*: *yang* derived from a character meaning *nourish*; *qi* = gas)
hydrogen (from Greek *water* + *brought forth*)	wodór (from *woda* = water)	Wasserstoff (*water* + *substance*; partially calqued on the Greek)	водород (*vodorod*; *water* + *origin*; calqued on the Greek)	氢气/氫氣 (*qingqi*: *qing* derived from a character meaning *light*; *qi* = gas)
nitrogen (from Greek *native soda* + *brought forth*)	azot (from Greek *without life*)	Stickstoff (from *smother* + *substance*)	азот (*azot*; from Greek *without life*)	氮气/氮氣 (*danqi*: *dan* derived from a character meaning *diluted*; *qi* = gas)

based on Greek). In general, Germanic and Slavonic terms tend to be more transparent than those of the Romance languages (and, to a lesser extent, English) in that they do not rely on Latin and Greek so consistently. Interestingly, in these instances, the Chinese words are less transparent than the previous examples: while each contains the character 气/氣, meaning 'gas', the precise meaning of the first character in each case may need to be hunted down in etymological dictionaries.

French and Italian were not included in the table as, etymologically speaking, in each case, the terms coincide with ones that have been included (French: *oxygène*, *hydrogène* and *azote*; Italian: *ossigeno*, *idrogeno* and *azoto*). The relative transparency of a term clearly has possible implications for the way in which the object or substance it represents is understood or interpreted, although one should be cautious about reaching hasty conclusions about such subtle, nuanced matters. However, it becomes a significant factor when it comes to identifying expressions as metaphorical, as the metaphoricity of the word *brontosaurus*, for example, seems to be of a different order from that of its Chinese equivalent 雷龙/雷龍.

1.2.4 *Scientific Writing*

Olohan (2016) presents a very thorough discussion of a large number of different kinds of scientific and technical translation, and her book includes chapters on both scientific research writing and popular science. One of the strengths of her treatment is the attention given to discoursal aspects of these genres. This is of relevance here of course, although it has to be said that this is more the case for some topics (e.g. mappings and, funnily enough, terminology) than for others.

Olohan highlights three particular features of scientific research writing. The first of these is the *move*. Moves are specific communicative functions (such as establishing or occupying a research niche), which can generally be correlated with identifiable text segments (2016:149–55). The second feature is *metadiscourse*, or, in other words, the devices that permit writers to organise text, engage their readers and indicate attitude (2016:155–7). Finally, *grammatical metaphor* refers to the mechanism whereby processes are frequently expressed as nominal phrases (2016:157–9). The first of these may have an application to metaphor in translation, but it is not explicitly investigated here, while the second is broadly reflected in the research contained in this book, principally in terms of the purpose parameter that forms the content of Section 5.1. Finally, grammatical metaphor is not touched on, as it is totally distinct from what is normally understood by the term *metaphor*.

While much of the aforementioned is probably reflected more or less equally in both research and popular scientific discourse, it is quite clear that there are significant differences between these two types of scientific communication, some of which are likely to have a bearing on the present discussion. Olohan summarises some of the main textual features of popular science reporting, while of the 15 items that she lists, the following find clear reflection in the texts that make up my data: emphasis on the novelty and importance of the content; tailoring of information to readers' existing knowledge and beliefs, avoidance of specialist terminology and explanation of terms that are included, direct quotation of scientists' words to enhance authority, presence of attitude markers to signal the writer's response or that which is expected of the reader, use of visuals to attract readers' attention and employment of similes to make the unfamiliar more familiar (2016:187–8).

Quite clearly, given the nature of my research, it is the last item that is of greatest potential interest, even though it focuses on simile rather than metaphor, as it provides independent confirmation of the use of figurative language in this genre of writing. In terms of the relative significance of this mode of expression in popular science and scientific research writing, Knudsen compares the numbers of metaphorical expressions in *Scientific American* and the more specialist journal *Science* and concludes that 3% of the words in the former are used in a metaphorical sense as opposed to only 1% in the latter (2003:1257).

Finally, Olohan considers how science news stories are localised or 'reframed' in translation in order to take the 'knowledge, interests and opinions' (2016:194) of the new audience into account. This process is reflected in the following issues: cultural specificity in terms of place names, institutions and other proper nouns, and measurements and currency; use of similes and comparisons with familiar concepts (although here I would want to include metaphor too); relevance of information for target readers; interaction with other public discourses, often of a topical nature (2016:194–9). The question of the *Scientific American* articles' further reading sections, which I briefly discuss in Section 2.3, comes under this heading; had it been a major goal of my research, this might have been extended into other areas such as front covers, images, and article titles and subheadings, as well as the relative acceptability to the different international readerships of cultural references, puns, simple analogies, inverted commas, question and exclamation marks, long (or short) sentences, technical (or general) vocabulary and innovative language.

1.3 Conclusion

In spite of some dissenting voices, metaphor is an inseparable part of scientific discourse and has been for many centuries. A small number of powerful metaphors are closely associated with particular areas of science, although as scientists' understanding develops, these can be either abandoned or adapted to enable them to continue to reflect the most up-to-date scientific thought. Metaphor plays a number of roles in science, some of the main ones being the channelling of scientific thought and the creation of new terminology. Olohan's analysis of scientific discourse is wide-ranging and nuanced, and includes many features that are of relevance to my study.

2 Translating *Scientific American*

This chapter is designed to provide a context for the discussion of translator decisions that will take up most of the rest of the book. After a section discussing the status of English as the international language of science, it will focus on the popular science journal *Scientific American* with the aim of exploring the background to the massive international translation effort that it is home to on a monthly basis.

2.1 English as the International Language of Science

Linguistically speaking, the period 1680–1970 was an anomalous one for science, according to Montgomery: during that time, scientific advances were not dependent on the existence of a single *lingua franca* (2013:134). Prior to that period, various *lingua francas* had been employed to communicate scientific ideas, while up until the end of the First World War, English, French and German were being used to approximately equal extents in scientific journals, with German subsequently declining quickly and French more slowly (2013:73). Throughout the century, this was a process that gradually gathered momentum, with the result that by the turn of the new millennium, English had effectively replaced national languages as the main vehicle for communicating scientific ideas (Montgomery 2013:134). This does not of course mean that there are not many other languages that have a fully developed scientific terminology; for example, although some borrowing of terms from English does take place, each of the six target languages considered here draws on its own resources to produce a scientific vocabulary that is largely independent of the English equivalents.

As will be described next, a massive monthly translation effort helps produce 14 international versions of *Scientific American* around the world. Along similar lines, the approach adopted by *National Geographic* has seen the appearance of nearly 40 international editions from the 1990s to today (Olohan 2016:192). By way of comparison, neither *Science* nor the *Lancet* is translated into any other language, while *Nature* has recently launched an Arabic version as its single foreign edition (Yahia 2012).

One of the chief explanations for why the popularising publications *Scientific American* and *National Geographic* are so widely promulgated while those such as *Science*, the *Lancet* or *Nature* are only translated very patchily is likely to be that, unlike the general public, scientists are by and large assumed to know English. Could it be that translation is becoming increasingly confined to popular rather than serious science? This is certainly a conclusion that appears to have been reached by Montgomery, who argues that texts that are translated out of English tend to be 'secondary works aiming at the dissemination of knowledge that has already been accepted by the relevant academic community' (2013:171). Bennett also seems to concur with this conclusion (2013:170).

However, if a knowledge of English can be assumed among scientists worldwide, then what is the position with the general public in the various countries where the editions we are studying are published? The Wikipedia article 'List of Countries by English-Speaking Population' (2016) conveniently summarises the percentage of English speakers by country and includes the following figures: Germany 56%, Poland 37%, France 36%, Italy 29%, Russia 5.48%. Individual sources are indicated for each language; while the percentage is stated to include both native and non-native speakers, the precise definitions used are likely to vary from country to country. Clearly, these figures refer to the population as a whole rather than just the (presumably well-educated) subsection that is likely to read popular science publications, although the proportionalities are likely to be broadly transferable. Taiwan is not included on the list, although the figure may well exceed the 0.8% that is quoted for the People's Republic of China.

2.2 *Scientific American* and Its International Profile

Writing in the April 1996 issue, *Scientific American* Editor-in-Chief John Rennie feels moved to remind his readers of magazine founder Rufus Porter's description of the publication as 'a paper that will instruct while it diverts or amuses' (Rennie 1996:6). As a non-refereed publication, *Scientific American* does not aim to publish original research so much as to provide accounts of recent work that has already been published elsewhere, written by the research scientists themselves for the readership. The magazine's audience probably comprises not only educated, although non-specialist, lay people but also some scientists who wish to keep informed about fields other than their own (Olohan 2016:174).

Most *Scientific American* authors are professional scientists describing their own work, with the periodical currently numbering 152 Nobel Prize winners among its contributors (Scientific American 2016c). Publication within the pages of *Scientific American* usually takes place relatively soon after the research has appeared elsewhere, which means that the work described within its pages is to all intents and purposes current. As a popular science publication, *Scientific American* aims to provide a bridge between technical science writing and a largely non-specialist audience, although clearly not one that is completely unversed in science. Although it does not shy away from conceptually complex material, it aims to make it more accessible through the use of vivid illustrations and an approachable writing style. Authors are encouraged to manage the task of drawing the lay audience into their subject almost in the fashion of a storyteller. The human context of the work needs to be described and technical concepts explained, but this often happens in a separate sidebar (SA Editors, personal communication, 2 February 2007). Interestingly, authors are also asked to bring to the fore any international aspects there might be to the work they are describing to ensure that their article will have the greatest possible appeal to the global audience to which it will be presented (Scientific American 2016c).

With a history of more than 170 years behind it, thanks to its status as an American national institution, the magazine has received numerous awards over the years for its excellent reporting of scientific issues both in its printed version and via its website (Scientific American 2016a). It has also been praised by luminaries such as Michael Bloomberg, Bill Gates, Al Gore, Jimmy Carter and Claude Lévi-Strauss (Scientific American 2016b), with Gore describing the publication as a 'national treasure' and Carter praising its 'unbiased presentation of the issues in this energy vs. wilderness debate' (Scientific American 2016b).

For the first half of 2016, the English-language edition reached an average monthly circulation of 399,354 (Alliance for Audited Media 2016—this figure is actually down by 70,000 as compared to that for the second half of 2013). It is read in print by around 3.5 million people worldwide (Scientific American 2016c), while the combined readership of the international editions totals more than 5 million (Scientific American 2016c). The ultimate answer to the question of why it is so widely circulated abroad, however, may have as much to do with U.S. cultural diplomacy and soft power as it does with profit margins and the desire to spread scientific knowledge around the world.

2.2.1 Scientific American International Editions

While *Scientific American* was founded in 1845, it was not until 1890 that the first issue of the Spanish-language *La América Científica e Industrial* appeared. This was the first foreign edition of the American periodical, and although it was only in existence for less than 20 years (Kanellos & Martell 2000:296), its publication marked the beginning of *Scientific American*'s international publishing endeavour. Its closure was followed by a period of some 60 years during which no international editions appeared, but from the late 1960s, the project of producing translated versions of the publication got underway once more with a vengeance.

The situation that has been reached today is that the journal appears every month in the following editions:

1. Italian: *Le Scienze* ('The Sciences')—1968
2. Japanese: 日経サイエンス (*Nikkei Saiensu*, 'Nikkei Science')—1971
3. Spanish: *Investigación y Ciencia* ('Investigation and science')—1976
4. French: *Pour la Science* ('For Science')—1977
5. German: *Spektrum der Wissenschaft* ('Spectrum of Science')—1978
6. Chinese (Simplified): 科学 (*Kexue*, 'Science')—1979; replaced by 环球科学 (*Huanqiu Kexue*, 'Global Science') from 2006
7. Russian: *В мире науки* (*V mire nauki*, 'In the World of Science')—1983–93, then 2002
8. Arabic: مجلة العلوم (*Al 'Oloum*, 'Science Magazine')—1986
9. Polish: *Świat nauki* ('The World of Science')—1991
10. Portuguese: *Scientific American Brasil* ('Scientific American Brazil')—2002
11. Chinese (Traditional): 科學人 (*Kexueren*, 'The Scientist')—2002
12. Czech: *Scientific American České vydání* ('Scientific American Czech Edition')—2002
13. Hebrew: סיינטיפיק אמריקן (Transliteration of 'Scientific American')—2002
14. Dutch: *Eos* ('Eos')—2003

The years of establishment are taken from Scientific American (2016d). Other editions have also come and gone over the years; for example, according to Archive.org, as of 4 August 2004, Greek, Korean, Romanian and Ukrainian editions were being produced. The possibility of unofficial translations into other languages appearing has not been investigated.

2.2.2 Structure of Editions

The translation of an entire journal issue is a complex event involving not only the conversion of textual material from one language to another but also more generally finalising the composition of the issue—which, in this instance, would involve decisions about what material from the source-language edition to include and the overall balance between translated and original material—and the construction and ongoing maintenance of a brand image within the new linguistic and cultural setting.

Just as today, back in 2004, a single issue of *Scientific American* consisted of many different elements. Besides at least half a dozen features—or in other words substantial articles, mostly written by scientists—space is also devoted to around 15 regular 'departments' (most or all written by *Scientific American* journalists) consisting of news, columns and reader contributions, and including items such as '50, 100 and 150 Years Ago', 'Anti Gravity' and 'Ask the Experts'.

Most or all of the aforementioned are written by *Scientific American* staff; not all of them feature in every issue. The feature articles, most of which are written by scientists rather than *Scientific American* journalists, take up about half the space in any given issue.

All the international versions that I have studied have a similarly complex structure: the main content is presented in the form of six to eight feature articles and the remaining parts comprise numerous columns and other shorter features. All the versions contain a mix of original and translated content, with the French and German versions weighted more towards the former and Italian, Polish, Russian and Chinese more towards the latter. In many instances, there is a greater tendency towards including home-grown content in the case of the shorter items, although the French, German and Italian editions do not even translate by any means all of the feature articles.

A range of approaches to timing and planning were indicated by the amount of time that elapsed (in 2004) between the publication of an article in English and the appearance of a translated version. At the one extreme, articles were invariably published in Polish and Chinese translation in *Świat nauki* and *Kexueren* one month after the appearance of the English original and almost invariably in Italian. For French and Russian, the time lag was around 1.5 and 3 months respectively on average. Finally, the German edition took by far the longest time to publish its translations, with the average delay being around 4.5 months.

2.3 Editorial Policy

While they were requested from each of the editorial offices, no style guides or internal instructions for the translators were made available for the various international editions. I therefore had no direct information regarding the editorial policies or projected readerships of the different editions. In spite of this, it was possible to infer some broad language-specific trends—such as a German tendency to paraphrase and reorganise and a French inclination to abridge—by studying macro- and micro-level changes to the texts. However, such specific observations took on less significance in the project as the emphasis gradually shifted from analysing the particular characteristics of each language version towards focusing on cross-lingual commonalities.

On the other hand, it was possible to obtain information from the editors of the French and German editions, so the very full responses that they provided to a wide-ranging set of questions that I asked are being presented in the belief that they are likely to be at least partially representative of all the international editions included in the study.[1] In both cases, the readership is (not surprisingly) perceived as being well educated and usually possessing a scientific background. Neither edition makes any particular recommendations to its translators in terms of the approach they should take. *Spektrum der Wissenschaft* provides feedback to individual translators where needed. Regarding the way in which the translation is carried out, *Pour la Science* states that their translators produce a literal version, but that during the editing process, the text can be changed, sometimes significantly. Clarifications can be added, the style can be improved, cuts can be made, or text boxes or illustrations added. The editorial line pursued by *Pour la Science* is different from that of *Scientific American*, as the French edition considers itself more demanding on a scientific level.

The French and German editions both prefer their translators to have a scientific background. Articles are allocated to translators according to speciality where possible and depending on availability. It would be reasonable to assume that this is the case for most or all the other international editions—or if not, at least this information provides some insight into the parameters along which variation is possible.

The decision as to which articles from an issue of *Scientific American* are translated is a particularly critical one for both *Pour la Science* and *Spektrum der Wissenschaft* given the selectivity with which both publications approach the English-language material. In making a decision, the French editors utilise the following criteria:

1. The interest of the subject matter
2. The author's status: articles not written by professional scientists are not usually accepted
3. The quality of the article
4. The international character of the article: one that is too American and cannot be transposed into the European context will be rejected

The editors of *Spektrum der Wissenschaft* read the articles that are in line with their own specialist areas and recommend either acceptance or rejection, with these recommendations being normally accepted without further discussion.

Editorial policy can also sometimes be inferred from the articles themselves. One example of this is the manner in which the further reading ('More to Explore') sections are handled. What emerges from a spot check of three articles is that *Kexueren* does not appear to have any kind of policy of offering target-language material to its readers, neither does *V mire nauki*. In other words, the three or four English-language items contained in the original section will be presented to the target-language readers in unmodified form, even though, ironically enough, these are the publications whose readers are probably least likely to be able to access the English-language material because of the low percentages of English speakers in the countries concerned. *Pour la Science*, *Le Scienze* and *Świat nauki*, on the other hand, do make some effort to supplement the original content with at least one item in the target language, although *Spektrum der Wissenschaft* is the only publication that seems to pursue a policy of always providing some target-language references. Given the fact that the international editions are (presumably) intended to make the English-language *Scientific American* material available to people who would otherwise be unable to read it, it is a matter of some surprise that the contents of these sections are not replaced by broadly equivalent target-language material at least partially on a more consistent basis across all international editions.

2.4 Translator Visibility

No information about the translators was available over and above what could be discovered from the pages of the magazines themselves. Exceptionally, each issue of *Le Scienze* gives the names of around three to five individual translators and attributes to them the particular articles they have worked on; in all, I found the names of nine translators listed across the several editions that I examined. *Kexueren* goes further, placing the name of the translator at the beginning of each article alongside that of the author; this publication draws on more than 20 translators, at least some of whom also write original content for the journal. However, these are the only language editions included in the project for which this information is made readily available. *Spektrum der Wissenschaft* lists between four and seven translators in each issue, although it gives no indication of how the various articles were apportioned; the editors identified these translators to me as an experienced group of staff members—a group that 10 years ago appeared to be greater than 40 in number. Just one single translator—the same person in each issue—is ever identified in *Świat nauki*, under the heading 'Translators, Authors and Consultants for the Current Issue', although because of the way in which the information is presented, it is quite possible that other people also contribute to the translation work. (Indeed, it is almost inconceivable that a single person could undertake the entire monthly translation effort single-handed.) No details about translators are supplied in *Pour la Science* or *V mire nauki*, although according to the information that they provided, the French editors work with only around five translators. Consequently, it is only possible to build up a partial picture of the numbers of translators working on the various language editions, with the information detailed earlier being the limit of our knowledge about the translators whose work forms the subject of the analysis sections that follow later in the book.

Although not all the figures are clear, the fact that relatively large numbers of translators are involved in most or all languages lends the data a greater level of generalising power than would otherwise be the case, in the sense that most of the findings of the research that follows cannot be attributed to the work of small numbers of translators.

2.5 Conclusion

With English now having acquired the status of an international language of science, there is less of a need for translating scientific research writing, with the result that translation effort largely focuses on popular science. *Scientific American* is a particularly outstanding example of how translation can be used to create a thoroughly international profile for a publication. The chapter has provided some insights into editorial policies, both stated and inferred, and has shown that none of the international editions is a straight cover-to-cover translation of the original. The numbers of translators involved in most or all international editions are such that the findings from the research that follows probably possess a good level of representativeness.

Note

1. I should like to thank Maurice Mashaal from *Pour la Science* and Hartwig Hanser from *Spektrum der Wissenschaft* for kindly providing answers to the questions I asked.

3 Metaphor and Translation

We didn't have metaphors in our day. We didn't beat about the bush.
(Fred Trueman, quoted in Fantoni 1998, unpaginated)

It was in the Upper Palaeolithic period, a recent theory suggests, that the human brain first became capable of metaphorical thought (Mithen 1996, 1998, 2005). According to this theory, up until that time, the domains that formed the main subjects of human thought were cognitively separated from each other so that no interaction or crossover between them was possible: quite simply, our ancestors at that time could not think about one thing in terms that were normally reserved for another (Kövecses 2005:24–5; Mithen 2005:233). During that period, however, there is ample evidence—in the form of artwork and tools that have been discovered—that human beings somehow acquired the ability to fertilise their thinking about one domain with elements taken from others (Mithen 1996:184–5, 2005:250–1). This in turn gave rise to modern humans' advanced symbol-manipulating skills, which arguably represent one of the most significant and characteristic features of the human cognitive system.

Whether or not this is true, it seems that metaphor is deeply engrained in the human cognitive system and to a large extent lies behind much of the symbolism, art and religion of which human beings have proven themselves capable over the millennia. It is also deeply embedded in language, the development of which Mithen identifies as the prerequisite for the phenomenal cognitive revolution referred to in the previous paragraph (2005:264): etymologically speaking, for example, many words in modern languages are based on components that derive from ancient languages or earlier stages in their own development. In addition—and this fact is crucial in the context of this research—metaphor is found in virtually every type of discourse, including scientific. Equally important for this research is the recognition that 'figurative language is one of the most productive sources of linguistic change' (Sadock 1993:44) and the realisation that 'most lexical items [are] dead metaphors' (Sadock 1993); we will find that the texts contained in our corpus are rich in expressions illustrating both these insights.

3.1 The Nature of Metaphorical Language

Tercedor Sánchez et al. describe the pervasiveness of different types of figurative language in specialised writing as 'a striking phenomenon' and one that is of great significance for term creation and the representation of technical and scientific knowledge (2012:71). This question was discussed briefly in Chapter One, but now it is time to turn our attention in more detail to the theoretical background to this research. In this fairly lengthy chapter, I will first consider the nature of metaphorical language. After that, I will discuss the subject that I prefer to call 'metaphor in translation', before finally moving on to a description of my research project.

3.1.1 What Is Metaphor and Where Does It Come From?

> Mother was looking more than usually puzzled. I knew that this was because she was not sure what a metaphor was. I told her in a quiet voice that a metaphor was describing something in terms of something else.
>
> 'Why should anyone want to do that?' she asked me.
>
> 'Either because you cannot find any other way to express what you mean, or in order to make a poetic or colourful effect. Like when you say God is love, or the cats are the very devil.'
>
> (Glendinning 1995:15–16)

Over the years, there has been considerable debate in metaphor research as to whether a metaphor constitutes a comparison, a substitution, a semantic tension, an interaction or a mapping. It is not the purpose of my research to contribute to this debate in any significant manner, although I do inevitably use one of these positions as a theoretical starting point. Metaphor is one of a bewildering constellation of related processes that involve some kind of transfer of meaning (some of the others being simile, analogy, synecdoche, metonymy, idiom, parable, irony, indirect speech act, euphemism, model and conceptual blend—to say nothing of translation, of course). Metaphor exemplifies the innate ability of the human mind to see or interpret one concept in terms of another.

To find the earliest theoretical discussion of metaphor, we need to go back to Aristotle in the fourth century BCE. Aristotle wrote about the subject in three of his works, *Poetics*, *Rhetoric* and *Topics*. Most writers have interpreted his position as viewing metaphors as 'implicit comparisons' whose use is 'primarily ornamental' (Ortony 1993b:3), although some take issue with this second contention, arguing that in his view 'good metaphors convey truths about similarities actually existing between things in the world' (Mahon 1999:76).

Moving forward more than two millennia, we find I. A. Richards in the 1930s putting forward his own form of what some contemporary theorists of metaphor have simply been known to dub the 'traditional view' (see, for example, Lakoff & Turner 1989:123). According to Richards, a metaphor consists of three components, the *tenor* (what the metaphor is about), the *vehicle* (the terms in which it is described) and the *ground* (the justification for this description) (1936:96, 117). Thus, for example, in the metaphor 'tiny cellular butlers known as transfer RNAs (tRNA) fetch the specified amino acids to be strung together' (Freeland & Hurst 2004:86) the tenor is the 'transfer RNAs', the vehicle the 'tiny cellular butlers' and the ground the fact that they 'fetch the specified amino acids'.

This is in a sense a 'classic' example of a metaphor, as all three elements are explicitly stated. In practice, however, it is more usual for only two out of the three to be present, with the third being supplied by the reader. This is the case with the metaphor 'Mitochondria, the tiny power generators within all kinds of cells' (Freeland & Hurst 2004:87), where the tenor is 'mitochondria' and the vehicle 'tiny power generators', but the ground—namely, that they generate a substance that provides most of a cell's energy supply—needs to be deduced by the reader from the context and from his or her knowledge of science.

Although Richards himself would have thought of a metaphor in terms of semantic tension between literal and figurative meaning (Ortony 1993b:3), in the aforementioned account, I deliberately used the relatively theory-neutral term *describe* to talk about the relationship between tenor and vehicle. Indeed, the fact is that it would be possible to use this model in such a neutral manner. The descriptive power of the approach lies in the fact that it enables one to identify and label the various components of a metaphor very clearly while also highlighting the contrast between literal and figurative meanings. However, although it has received some development, compared with a number of other approaches, the metalinguistic apparatus of this theory has not been elaborated to any significant extent.

Sadly, there is no space for a detailed discussion of possible approaches to the study of metaphor. The decision has been taken to use the so-called cognitive or conceptual metaphor theory largely because of its metalinguistic apparatus, which is both sophisticated and convenient, and because of the way in which it facilitates a multi-parameter analysis of metaphor in text. A side product of this research is that it presents an opportunity to test the validity of some concepts that originate within the conceptual metaphor theory, although this is by no means its main aim. At the same time, of course, much of the content—such as Section 4.1 and Chapter Five—employs concepts that do not generally form part of this approach but have not as yet been theorised to any substantial degree.

A theory that provides the versatility, flexibility and intellectual rigour to account for real-world metaphorical phenomena of the kind that the book will be examining will need to encompass and provide explanations for a wide range of observations. The list of such topics should ideally include most of the following: metaphor and its range of applicability; creative, conventional and dead metaphor; metaphor and idiom; metaphor structure; literal versus figurative meaning; types of figurative meaning and figurative expression; functions of metaphor; metaphor, etymology and polysemy; core and peripheral meaning; metaphor, neologism and borrowing; patterns of metaphor in text; and metaphor and meaning components (see Knowles & Moon 2006:1–22).

3.1.2 The Conceptual Metaphor Theory

By far the most influential framework for the study of metaphor is currently provided by the conceptual metaphor theory (which is also known as the cognitive theory of metaphor). The American scholars George Lakoff and Mark Johnson were not the first to present the idea of an unspoken guiding metaphor underlying the metaphorical expressions that exist in a text (see, for example, Keesing 1985:216–7, Jäkel 1999 and Dobrovol'skij & Piirainen 2005:121). However, since 1980, when the idea was first proposed in its current form by Lakoff and Johnson, it has been considerably developed both by these original two authors (Johnson 1987, 2007; Lakoff 1987a, 1993; Lakoff & Turner 1989; Lakoff & Johnson 1999) and also by scholars such as Turner (1991, 1997), Fauconnier (1994), Gibbs (1994), Fauconnier and Turner (2002) and Kövecses (2002, 2005, 2006, 2010).

The conceptual metaphor theory is designed to counter what Lakoff identifies as five false assumptions about literal language. Of these, the first three are perhaps of greatest relevance for us now:

1. All everyday conventional language is literal, and none is metaphorical.
2. All subject matter can be comprehended literally, without metaphor.
3. Only literal language can be contingently true or false.

(1993:247)

What emerges from Lakoff's position is a view of metaphor as an integral part of language and certainly not simply as an ornament or embellishment. Consequently, a metaphor should be thought of not as an isolated figurative expression in a text, but as a 'mapping' from one domain of experience (the 'source domain') to another (the 'target domain'), which is a kind of mental connection made between two unrelated concepts or areas of experience that allows one to think and talk about one of these concepts or areas in terms usually reserved for the other and that potentially sanctions an unlimited number of individual metaphorical expressions relating to that particular area (Lakoff 1993:203; Evans & Green 2006:295).

I will be using this term 'mapping' in preference to 'conceptual metaphor' or 'underlying metaphor' for the reason that I state in Section 4.2.1. A mapping—a concept that many authors refer to as a conceptual metaphor—can be highly specific to a particular context or even a single word (for example, the word *sheathing* in the expression 'when to start *sheathing* those axons'

(Fields 2004:59), which implies the mapping AN AXON IS A SWORD), in which case it would be unlikely to engender many additional metaphorical expressions, or else it can be more productive (for example, 'neuronal *circuits* that are not *wired together* directly' (Fields 2004:61) and 'the brain's reward system: *a complex circuit* of nerve cells' (Nestler & Malenka 2004:78, 80), which are just 2 out of 12 different expressions in my data that are derived from the mapping THE NERVOUS SYSTEM IS CIRCUITRY). The distinction between macro-level mappings (i.e. conceptual metaphors) and micro-level metaphorical expressions will be of central importance in what follows. In addition, a hierarchical relationship often exists between mappings, whereby the source domain of one is hyponymic to that of another (e.g. NATURE IS A SOFTWARE DESIGNER and NATURE IS A HUMAN, where *software designer* is a kind of hyponym of *human*).

Within this view, metaphor is generally seen as a means of explaining difficult, unusual or abstract concepts. A metaphorical expression such as 'biological approaches to mental illness gradually *overshadowed* psychoanalysis' (Solms 2004:84), for example, depends for its meaning on seeing the relatively more complex concept of 'biological' approach as if it were a large object that was blocking the light to a smaller object. Consequently, the less familiar is generally modelled in terms of the more familiar; thus in the metaphor 'unneeded synapses get *pruned*' (Fields 2004:61), neurons are depicted as trees or plants. In this way, a metaphor can act as a stepping stone towards understanding a difficult concept. Also, the more abstract concept is modelled in terms of the less abstract. For example, in the phrase 'two artificial genetic *"letters"*' (Gibbs 2004:76), genetic material is portrayed as if it were text.[1]

Interestingly, the fact that this mapping for genetic material exists does not preclude other ones from also being used in parallel, and indeed, other examples within the data model this concept in terms of code, instructions or software, for instance. Each member of this 'cluster' of mappings highlights certain characteristics possessed by genetic material and conceals others depending on the author's particular communicative purpose. In addition, a mapping tends to be partial in nature (Kövecses 2010:103), as in focusing on particular characteristics of the target domain other aspects of it generally remain undeveloped. This is how it is possible for eight separate mappings for genetic material (in which it is seen as circuitry, code, data, instructions, language, machinery, software or text) to exist alongside each other. Finally, a mapping may be exploited in both conventionalised and more original expressions. For example, the mapping GENETIC MATERIAL IS SOFTWARE makes possible a number of conventionalised expressions (e.g. 'The aim . . . was to *reprogram* cells': Gibbs 2004:78) as well as the occasional more creative one (e.g. 'think of it as *Life, version 2.0*': Gibbs 2004:76).

A consequence of the conceptual view of metaphor is the relegation of the notion of 'dead metaphor' to a mere result of the (in their terms, mistaken) objectivist view of meaning and understanding: rather than being 'dead', a word that derives its fixed meaning in a metaphorical manner is simply a single example of a particular metaphorical structure that forms a small element of the expressive resources of a given language (Lakoff & Johnson 1980:211ff.). By a similar token, the idea that such an expression 'was once a metaphor' (1980:212) is also rejected, as is the view that the fossilisation of once live metaphors is 'a typical way of providing words for preexisting meanings that lack words to express them' (1980:213).

While most metaphors conform to the pattern described earlier, one category, known as image metaphors, does not. Relatively little has been written on this type of metaphor (see, for example, Lakoff 1987b; Caballero 2002, 2003a, 2003b), in which the mapping is not between domains but between images (e.g. portraying a bulbous axon terminal as if it were a bud, or a cell-free void separating two areas of frenetic neuron activity as if it were a highway dividing two areas of burning forest). In other words, an image metaphor's underlying mapping is generally based on similarities of appearance, even if this is across vast differences of scale, for example. Many, but by no means all, image metaphors can be described as 'one-shot': highly specific expressions that only fit in one specific context.

One matter that the conceptual metaphor theory does not touch on is the counter-tendency for metaphors to be used to conceal rather than to reveal, or to mystify rather than to clarify. This idea is present as far back as Aristotle who in his *Poetics* (XXII) presented the view of metaphors as riddles. Intuitively, this represents a significant use of metaphorical language, and is certainly what is generally alluded to by an injunction not to 'talk in metaphors'. Be that as it may, this metaphorical usage is not encountered in this present study: it would indeed be strange if metaphorical expressions were used in this manner in scientific discourse.

3.1.2.1 Applying the Theory

The concept of mappings is certainly an interesting and compelling one, and the possible implications that it has for translation are certainly far-reaching, both on a theoretical level and in terms of how the problems they might present to translators could be overcome. However, Lakoff and Johnson themselves do not investigate languages other than English, while the premise of some writers has been that the mappings that can be commonly observed in English are for the greater part universal in nature. For the same reason, most important monographs on metaphor have tended not to discuss the question of translation. Because of this, the conceptual metaphor theory in the form in which it is presented by Lakoff and Johnson is perhaps not the best theoretical framework for the study of metaphor in translation. With much of its work based on the analysis of a single language, it not surprisingly tends to downplay the amount of interlingual variation in metaphorical patterns that is caused by linguistic or cultural factors—a position that arguably became more pronounced over the first 20 years of the theory's development (see Engstrøm 2000:268). Instead of such factors, the emphasis is generally placed on what is termed *embodiment*, or in other words what is argued to be our cognitive system's dependence on our bodily experiences in forming the categories by means of which we make sense of the world around us (Kövecses 2010:107–20). One of the theory's main tenets is consequently somewhat at odds with some of the conclusions that have been reached by translation scholars over the last two decades and more, the result being that some conceptual metaphor theorists emphasise the universality of most metaphor, whereas the notion of interlingual and intercultural variation, which is part of the lifeblood of translation studies, is often downplayed.

Generally speaking, the fact is that we know much less about the metaphorical structures of non-English languages than we do about those of English. However, this need not be a serious problem for us here: as is going to be discussed in Section 3.2.1.1, translation studies concerns itself with spontaneous, non-systematic and only semi-predictable language use of the type described by Saussure's concept of *parole*, and in addition, even if we know about the metaphorical structures native to all the target languages in great detail, that will only be of partial assistance in view of the interaction with source-language structures that would inevitably occur.

It is certainly true that there seems to be an amount of overlap between languages when it comes to some of the more major mappings (e.g. ARGUMENT IS WAR, HAPPY IS UP), although there is no conclusive evidence as to the universality even of such apparently fundamental mappings, and it also appears that the way in which every mapping manifests itself is subject to a certain amount of interlingual and intercultural variation. In an attempt to pin down this variation, in one of his two books about metaphor and culture (2005, 2006) Kövecses discusses, for example, how different components of a conceptual metaphor—such as the source and target domains and how they relate to each other, or the particular mappings or specific entailments involved—can be involved in variation across languages and cultures (2005:117–30). Apart from this scholar, many other metaphor researchers and cognitive linguistics specialists utilise at least some of the concepts and methodologies of conceptual metaphor theory while at the same time recognising that language and culture are vital factors in determining metaphor usage—for example, Deignan, Gabrys and Solska (1997); Gibbs (1999a, 1999b); Semino (2002); Charteris-Black (2003); Dobrovol'skij and Piirainen (2005); Caballero and Díaz-Vera (2013); Musolff, MacArthur and Pagani (2014). Deignan argues that taking account of cross-linguistic differences is not incompatible with the conceptual metaphor theory (2005:100), although it is likely that in this respect, Deignan and her like-minded colleagues envisage a higher level of interlingual variation than would writers such as Lakoff and Johnson.

Another potential problem with the approach adopted by Lakoff and Johnson from the point of view of translation studies research is their repeated reuse of the same made-up examples. For instance, the amount of overlap between the examples used to illustrate the LOVE IS A JOURNEY mapping in significant works of theirs published at an interval of virtually 20 years is striking, as the following two lists indicate.

First, their 1980 work *Metaphors We Live By* gives a list of 13 expressions illustrating this mapping that includes the following:

> Look *how far we've come.*
> We're *at a crossroads.*
> We'll just have to *go our separate ways.*
> We can't *turn back now.*
> I don't think this relationship is *going anywhere.*
>
> (1980:44–5)

In *Philosophy in the Flesh: The Embodied Mind and Its Challenge to Western Thought*, another major co-authored work published in 1999, a slightly different list of 13 expressions illustrating the same mapping is provided, which includes the following examples:

> Look how *far* we've *come.*
> We're at a *crossroads.*
> We may have to *go our separate ways.*
> We can't *turn back* now.
> The relationship is *not going anywhere.*
>
> (1999:64)

Apart from a couple of minor rewordings, the only difference is a slightly more nuanced approach to the use of italics to indicate words that are being used metaphorically. Besides these five items, in each list there are four further expressions that overlap with each other, and four that are different. In other words, readers are expected to derive their full understanding of this mapping from a static set of introspection-based examples rather than being presented with a dynamically developing collection of real examples that reflect how the mapping is used in practice.

In contrast to this reliance on non-authentic, made-up examples, there are in fact many studies that are based on real data and that make use of corpora (see, for example, Charteris-Black 2004, 2005, 2013; Deignan 2005, 2008; Gries & Stefanowitsch 2006; Stefanowitsch & Gries 2006; Deignan & Semino 2010; Low, Todd, Deignan & Cameron 2010; Nasti 2012). The advantage of these is that, being based on analyses of metaphor derived from real data, they can offer much more objective insights into a wide range of different language patterns.

Dobrovol'skij and Piirainen, two researchers who take very seriously the need to base research on the use of real data and to take full account of cultural variation, provide the following statement regarding their use of the conceptual metaphor theory:

> The Cognitive Theory of Metaphor is important for investigations into the phenomenon of figurativeness because it provides the researcher with a well-developed metalinguistic apparatus, including heuristically significant concepts such as source domain, target domain, metaphoric model, conceptual mapping, conceptual correspondences, metaphorical entailment, etc. In many cases, applying this apparatus allows us to explain many real properties of figurative units which could not be captured in the framework of any traditional approach.
>
> (2005:142)

While they base many aspects of their research on the conceptual metaphor theory, they modify and adapt it wherever they deem this necessary in order to take account of their data. It is this way of applying the theory that I have adopted in my work.

3.1.3 Categorising Metaphors and Metaphorical Expressions

The dimensions along which metaphor can be classified are almost unlimited. A number of those discussed in what follows have undergone significant theoretical development in the literature; others, however, have received considerably less attention. The parameters that are discussed below will form a major part of the analytical apparatus used in this research. The discussion in this chapter is of a general nature, and the concepts introduced here will be adapted and contextualised for use within my research.

In the sense in which the term is used in this research, a parameter is taken to mean a feature of metaphorical language (such as subject area, degree of novelty or level of cognitive load) that may vary from metaphor to metaphor or metaphorical expression to metaphorical expression and that may, conceivably, affect the way in which a particular metaphorical item is translated. It is this possible influence that is one of the topics that my research is intended to investigate. All six parameters considered in this work are derived from metaphor research.

To initiate our discussion of parameters, let us look at four kinds of metaphor identified by Lakoff:

1. Complex schema mappings (e.g. ARGUMENT IS WAR)—i.e. conceptual metaphors based on propositional knowledge (see Section 6.1.1.2)
2. Image-schema mappings (e.g. CONTAINERS, PATHS, LINEAR SCALES and CENTER-PERIPHERY): like 1. in this list, these are conceptual metaphors, but are based on image schemas (see Section 6.1.1.1)

3. One-shot, rich-image mappings (e.g. 'the Schwann cells *sprouted branches*'): i.e. image metaphors (see Section 6.2.1)
4. Aristotle's metaphor (e.g. *Man is a wolf*, *Harry is a pig*)—i.e. SOMETHING IS WHAT IT HAS SALIENT PROPERTIES OF (not covered in any section) (1987c:194–5)

According to Lakoff, 'In general, useful scientific analogies seem to have the properties of metaphors of Type 1; that is, they are systematic, well-defined, and map high-level abstract relations such as causation' (1987c:196). Expressions conforming to the pattern of Aristotle's metaphor tend to be largely restricted to the pages of textbooks on metaphor and rhetoric, and unsurprisingly there are no examples in my data. Significantly, however, there is a major exception to this, which is that most or all mapping verbalisations are likely to be examples of this type. (Additionally, the epigraph to Section 3.1.1 also contains two instances.) Taken together, these four non-mutually-exclusive categories (or the first three, at least) form the basis for the provenance and richness parameters that I will introduce below.

It would be wrong to imagine that the six parameters that I use in my research are the only ones that might have been selected for a study such as this. The number of dimensions along which it would be possible to categorise metaphor is probably open-ended. Indeed, different authors focus on different parameters in their work. For example, Cameron, a metaphor scholar rather than a researcher into metaphor in translation, discusses the following nine 'graded descriptors of metaphor' (1999:123) for the purposes of identifying and describing metaphor in spoken discourse data:

1. Degree of incongruity between topic and vehicle
2. Novelty/conventionality of topic-vehicle link
3. Attitudinal impact
4. Explication
5. Familiarity
6. Cognitive demand of topic and vehicle terms and domains
7. Explicitness of metaphor: the receiver's conscious awareness of the producer's metaphorical intention
8. Connotative power of the vehicle system
9. Systematicity: the extent to which the same metaphor or variations on it are used in discourse, locally or globally

(1999:124–30)

Besides these, there are many more dimensions for classifying metaphors that could potentially be exploited in this kind of research. These would include basic level domains (Rosch 1978; Evans & Green 2006); conceptual blending (Grady, Oakley & Coulson 1999; Fauconnier & Turner 2002); single versus multiple metaphor, simple versus complex metaphor, pure versus mixed metaphor (Schaeffner 2012; Gibbs 2016) and restricted versus extended metaphor (Crisp, Heywood & Steen 2002:60–2); implicit versus explicit metaphor (Steen 1999:61–4); nominal, predicative and sentential metaphor (Miller 1993:382–6); and the linguistic forms in which metaphor appears (Brooke-Rose 1958; Goatly 1997; Deignan 2005:145–68). This last parameter might also include metaphor and collocation (Deignan 2005:-193–213), the presence or absence of a signalling device (such as inverted commas: see Section 6.1.3.1.3 point v)), the part of speech of the principal component of a metaphorical expression and the presence or absence of topic, vehicle or ground.

It would also be possible to focus on parameters taken from translation studies—for example, levels of explicitation, simplification or cultural embeddedness—but since this work sets out to explore the interface between the two disciplines, I decided not to pursue this approach. Looking at variations between the outputs of different translators would have been implementable for Italian and Chinese, but not for the other languages since no precise information (and, in the case of some languages, no information at all) was available for them on the identity of the translators (as discussed in Section 2.4). It was in view of this lack of information, and because the work concerns itself with the commonalities of translation practice that can be observed across languages, that this particular angle was not pursued. Furthermore, in a sense, the anonymity of the translators in most language editions takes on less significance by dint of the number of translators involved: the theory is that by taking all the target languages together we will simultaneously be looking at the output of up to six translators for each English example analysed so that the effect of any unusual or idiosyncratic decisions by individual translators will be significantly diluted.

Translation scholars have also utilised a range of different metaphor-based parameters. Dickins' 'Full Model', for example, involves the use of a total of six 'dimensions':

1. Topic, vehicle, sense/grounds
2. Lexicalisation versus non-lexicalisation
3. Schematicity
4. Reflected meaning
5. Perceived recency of metaphor
6. Technicality versus non-technicality (2005:265)

As with those of Dickins, the motivation for the particular set of parameters that I have selected is 'textual analysis of metaphor in a translation context' (2005:227), although having employed a range of theoretical concepts taken from metaphor research, an additional concern of mine is to be able to offer something back to that discipline. As mentioned in the introduction, the six parameters that my research focuses on are existence category, mapping, purpose, conventionality, provenance and richness, and the inclusion of each will be justified in the appropriate chapter. The notion of cultural embeddedness—clearly an important one for the study of metaphor in translation—emerges from a number of these, such as conventionality (see Section 5.2.1) and richness (Section 6.1.1.3), as the discussion proceeds.

3.2 Metaphor and Translators

Metaphor is perhaps being considered less and less of a 'problem' for translators, but it is and always will be a textual feature that needs to be dealt with along with many other textual features. Apparently, translators are able to take many metaphorical expressions in their stride, while others do appear to give rise to some serious difficulties. What is special about metaphor, however, is the fact that it has a conceptual dimension, is linked to varying degrees with cognition and, in this way, is closely connected to the ideology, beliefs or, in this case, knowledge systems that inform a text. Also, the fact that metaphorical expressions frequently do not occur in isolation means that the dimension of coherence also needs to be taken into account. In many cases, rather than being deeply problematic, it is simply one textual element—albeit a significant one—that needs to be handled in a manner that is appropriate for the translation's intended audience.

3.2.1 *Introduction and Overview*

It was in 1965 that Catford published his now famous definition of translation: a formulation that depicted what the translator does as 'the replacement of textual material in one language (SL) by equivalent textual material in another language (TL)' (1965:20).[2] This definition, and the theory of which it formed a part, were perhaps typical of the tendency in those days to try to reduce translation to a purely linguistic phenomenon, and Catford devoted much serious thought to working out a theory that was intended to account for all aspects of translation in the somewhat mechanistic terms of his definition. Linguistics has, of course, moved on considerably since the 1960s, while in the 1980s, a 'cultural turn' (see, for example, Malmkjær 2005:36; Snell-Hornby 2006:47–68) took place in translation studies that introduced many new factors that are now accepted as being of central importance to the study of translation. Research into metaphor in translation has quite clearly benefited from this, even if a more significant influence on its development has perhaps been played by metaphor research itself.

3.2.1.1 *Translation Studies: Portrait of a Discipline*[3]

The discipline that is now known as translation studies properly originated in about the 1960s. Up until that time, the study of translation had been largely non-systematic and sporadic, and it had focused largely on normative questions of quality, accuracy, translatability and the correct way to translate, almost invariably with reference to either literary or religious texts.

Much has changed in the last 40–50 years, however, and translation studies has formed itself into a fully fledged discipline, complete with theoretical paradigms and a non-prescriptive approach to the investigation of translation. In line with this, translation is now seen as an act of communication situated within a particular linguistic and cultural context, and influenced, for example, by genre conventions, audience expectations and the translator's own interventions in the text, whether these be unknown or intentional (see Hermans 1985:11). This is indeed a very different—and more highly nuanced—vision from what was generally current in the 1960s and what lay behind Catford's definition cited earlier. Indeed, all of these new insights are concepts that are glaringly absent (as it now appears) from most earlier approaches such as that of Catford.

Another important point is that translation studies is not simply contrastive linguistics by another name, nor is it (any longer) a mere branch of applied linguistics. To understand one of the main reasons why translation studies is fundamentally different from contrastive linguistics, we must look at a distinction made by the early twentieth-century Swiss linguist Saussure. According to Saussure, within the general concept of language, it is necessary to distinguish two separate notions, those of *langue* and *parole* (Saussure 1986:9–10, 15; see also Holdcroft 1991:19–46; Koller 1992:222–3). This is a distinction that was highly influential in twentieth-century linguistics. *Langue* represents the theoretical, abstract concept of a language as a rule-governed system, while *parole* denotes instances of specific language usage that are produced in response to a particular set of stimuli. Thus, for example, formal grammar, lexical semantics and contrastive linguistics are predicated on *langue*, while discourse analysis and pragmatics take *parole* as their main object of study. Translation studies, which focuses on analysing real acts of communication and language use within authentic contexts, is also a discipline that is primarily dependent on the study of *parole*, as well as of translation norms (see Hermans 1999; Schaeffner 1999), a third concept that allows researchers to organise parole-based phenomena into recognisable behaviours and patterns. Metaphor research arguably concerns itself with both, but if my understanding is correct, it is ultimately interested in building up our insight into how metaphor functions as part of a language system (i.e. *langue*) and, indeed, of human psychology. This fundamental distinction is therefore important for understanding the different focuses of these two disciplines.

The reason translation studies cannot be considered a branch of applied linguistics is that it is an area of study that has been deeply influenced by a wide range of different disciplines. The list of such disciplines is potentially long; Munday, for example, includes various branches of linguistics, modern languages and language studies, comparative literature, cultural studies, philosophy of language, sociology, history and creative writing as having exerted a significant influence (2012:24–5), but in fact this list could probably be extended still further. In view of the highly interdisciplinary nature of modern translation studies, caused by the extremely wide range of other disciplines that have fed into it, it has even become customary to talk of it as an 'interdiscipline' (a trend that seems to have been set in motion by Snell-Hornby, Pöchhacker and Kaindl 1994).

Metaphor is, of course, only one of many possible objects of research within translation studies. In a highly programmatic article originating from the early 1970s that has played a very significant role in lending the discipline its current shape, Holmes defines translation studies as consisting of three branches: the theoretical, the descriptive and the applied (2004:184–91). Holmes's overall vision of the discipline is often presented visually as a map (see, for example, Toury 1995:10). While purely theoretical work is certainly being carried out—either on a 'general' all-encompassing theory or on more specific partial ones (2004:186)—most of the research that is not of an applied nature takes place within the descriptive branch, the aim of which is to investigate translation as both a product and a process, and also the functions that it possesses, within a given linguistic, cultural, social or political context (2004:184–5). The term 'descriptive translation studies' (along with the name of the discipline itself, selected in preference to alternatives such as 'translation theory' or the 'science of translation': 2004:182–3) implies a non-prescriptive approach to researching a wide range of topics firmly oriented towards the target rather than the source (Toury 1995:23–39) and committed to the analysis of real instances of translation. Furthermore, the cultural turn has engendered a greater diversification in the discipline, which now largely concerns itself, for example, with translation norms, rewriting and manipulation (Lefevere 1992), postcolonial contexts (Niranjana 1992; Robinson 1997), gender issues (Simon 1996; von Flotow 1997, 2010; Santaemilia 2005), questions of translators' visibility (Venuti 2008) in the texts they produce and the application of narrative theory (Baker 2006) to acts of translation. Holmes's map has been extended too, most notably perhaps by van Doorslaer (2007).

Interestingly, many valuable insights into translation have been gained through the deliberate exploitation of old metaphors for translation, or the deliberate introduction and exploration of new ones. St. André (2010) offers a collection of articles that present a wide range of these (see, for example, Tyulenev 2010 for translation as smuggling; see also Hönig 1997b for translation as bridge building; Johnston 1999 for translation as simulacrum; Vieira 1999 for translation as cannibalism and vampirism; Hermans 2007 for translation as transubstantiation; Evans 1998 and Guldin 2016 for general discussions).

In spite of all the new directions for research outlined earlier, there is still plenty of scope for text-based investigations that focus chiefly on matters of language. What, then, are the parameters that define this continuing activity in terms of both the areas studied and the methodology?

There are in fact a wide range of different areas that form the objects of research. However, instead of translated material being analysed against a particular 'correctness notion' (Hermans 1991:166), translation is seen as a process of decision-making that leads to both loss and gain. There are no 'right' or 'wrong' translations—just degrees of appropriateness. Mistranslations, mistakes and standard translation equivalents are not of significant interest to the discipline. In line with this, a move has occurred away from focusing on translation shifts—that is, minor rewordings—that are caused by small incompatibilities of grammar and lexis, which is the situation that we see presented in Catford's understanding of shifts (1965:73–82) and that was presumably caused by the influence of paradigms derived from contrastive linguistics. In point of fact, all of Catford's shift types relate to grammatical matters over which translators have no significant control, such as the position of an adjective relative to its noun or the order in which subject, verb and object typically appear in a sentence (1965:73–82), and for this reason, they are not of any significant interest to translation scholars today. Instead, what we see now is a focus on shifts that are optional and that reflect conscious decisions made by translators in an attempt to produce a viable solution to a real translation problem. In other words, what is now often of interest is the target context: what a translator chose to say as opposed to what could have been said.

If shifts are located on the micro-level (i.e. that of individual words and phrases as they exist within a text), then there is also much interest within the discipline in what occurs on the macro-level (the level that involves larger textual structures). Translation studies no longer places its exclusive attention on studying single words, phrases and sentences in isolation but now also examines what happens to higher-level entities—such as an extended section of discourse, a whole metaphorical structure, a line of narrative, an authorial voice or an entire text—when translated into a different language.

As will be discussed in Section 3.2.2.5, modern text-based translation studies tends to distinguish between strategies and procedures. However, care has to be taken not to attribute specific internal motivations for particular actions to a translator purely on the basis of external—i.e. textual—evidence. A further notion, that of translation effects, can also sometimes be used, although this is less common. These would refer to the consequences of a translator's decisions, rather than the techniques used in order to achieve a satisfactory target-language version. For the sake of simplicity, this research does not distinguish rigidly between shifts, effects and procedures, with the result that some processes that are labelled as 'procedures' will have a higher degree of intentionality about them than others.

As regards specific topics selected for investigation within this type of research, besides the investigation of how translations differ from their source texts, considerable attention is also paid to the nature of translated texts, how they differ from original texts written in the same language (see, for example, Toury 1995:207–17) and how the language of translated texts can in general be characterised.

Within this context, research into the so-called universals of translation does continue, even if there is much less of a tendency to use this controversial term (see, for example, Mauranen 2004, 2008; Ilisei 2012). Here it is worth pointing out that the most widely researched of these is probably explicitation (or in other words 'the phenomenon which frequently leads to TT stating ST information in a more explicit form than the original': Shuttleworth & Cowie 1997:55; see also Pápai 2004; Becher 2011). Besides this, translation scholars investigate a wide range of textual phenomena: not only metaphor, of course, but also punctuation, lexis, collocation, fixed expressions, word play, cultural realia, allusions and semantic prosody (for this last item see Malmkjær 2005:115, 130–1; Munday 2011). Some features are of course situated on the macro-level, such as, for example, non-standard discourse or pronunciation, intertemporal translation, implicature and questions of genre (see, for example, Hatim 1996; Hatim & Mason 1997; Malmkjær 2005).

One of the most popular methodologies is undoubtedly the corpus-based approach. Even where this is not appropriate, research is still almost certain to reflect translation studies' healthy emphasis on the use of real data. Investigations can be either qualitative or quantitative, or a combination of the two. Every bi- or multilingual study generally starts with a monolingual one (i.e. an investigation of the source material). This is generally followed by text alignment or the matching of individual source and target items (i.e. source-text problems and their target-language solutions) for the purposes of analysis. Alternatives to this paradigm for bi- or multilingual research have not yet been seriously investigated.

3.2.1.2 Translation Studies: Its Relevance to Metaphor Research[4]

In spite of the interesting work that was being conducted by translation scholars within their own discipline, in the mid-1990s Mandelblit stated that the treatment of metaphor in translation studies was 'thoroughly at odds with the findings of the Cognitive Linguistics research on metaphor' (1995:485). Whether or not this was a totally fair assessment, from around that time at least, some scholars interested in metaphor in translation have sought to draw increasingly systematically on the insights of this other discipline, recognising the great scope for further interaction that exists between the two disciplines.

With this in mind, it should be expected that any proper research into metaphor in translation that has absorbed significant insights from metaphor research should be in a position to feed findings and results back into that discipline, even if that is not the primary aim of the research and such findings and results are only a kind of by-product of the research. It is true, however, that some of the claims made by translation scholars as to the centrality of translation to the study of metaphor will probably strike most metaphor scholars as somewhat overstated. For example, Dagut declares that 'it is translation theory that holds the key to a deeper understanding of metaphor' (1976:32), and Kurth also makes an assertion that is very similar (1995:120). On the other hand, Guldin lists three 'points of contact' between the two disciplines: the use of specific metaphors to describe how translation functions, the exploitation of the concept of translation as a metaphor for exchange and transformation in different types of discourse and the question of developing procedures for the translation of metaphors (2010:161–2; see also 2016). To these he adds a fourth, that of the structural similarities and etymological parallels that exist between the concepts of translation and metaphor (2010:162), as was discussed in Chapter One.

In addition to Guldin's valid observations, it is likely that translation studies can furnish metaphor scholars with evidence of the vast numbers of discrepancies that exist between languages on the micro-level (and macro-level, for that matter). Indeed, translation studies can provide many case studies—mono-, bi- and multilingual—to supplement the descriptive work of metaphor scholars. (Along similar lines, Malmkjær argues that translation studies can be a source of data for linguistics: 2005:58–60). More ambitiously, however, Mandelblit argues that translation examples 'can provide excellent data for studying the "hidden" cognitive processes involved in language usage' (1997:20).

Whether or not this is the case, translation scholars are generally not only bilingual but also bicultural, and they can provide an excellent source of data regarding interlingual and intercultural variation to counterbalance the universalist tendencies of the conceptual metaphor theory that still survive in some parts of the discipline. However, it needs to be borne in mind that the kind of data that it offers is fundamentally different from that supplied, for example, by a *langue*-based discipline such as contrastive linguistics. This is because translation studies data cannot be relied on to reflect target language norms in an unbiased manner, contaminated as it is by the influence of the source language, by the translator's preference for a particular set of solutions and so on, all of which factors place this data firmly in the domain of *parole*, as argued in Section 3.2.1.1. In other words, as instances of mediated discourse, examples of translated metaphorical expressions cannot generally be used as primary data for metaphor research.

That said, translation-related data can potentially serve as a test bed for measuring the validity of some metaphor research categories and concepts: one may reasonably argue that, if certain factors influence speakers of a particular language (or language in general), then they must do the same for translators as they make decisions regarding wording and so forth—whether such factors are universal in nature or bound to a particular language.

3.2.1.3 Metaphor in Translation: An Introduction

Since the discussion of metaphor in translation was initiated 50 years ago by Kloepfer (1967) much has been written on the topic and a number of important debates have been pursued. One of the earliest of these concerned the extent to which metaphor constituted a problem for translators (whether there was 'no problem' or 'no solution': see Dagut 1976:25 and Section 3.2.2.1) and the fallout from Kloepfer's controversial remark that 'the bolder and more creative the metaphor, the easier it is to repeat it in other languages' (1967:116; translation taken from Snell-Hornby 1995:57). Besides this, many writers have proposed lists of procedures for translating metaphor (e.g. reproducing the original image, replacing the metaphor with a simile, converting it to sense; see, for example, Newmark 1985:304–311, quoted in Section 3.2.5). Next, there has been a gradually awakening awareness that metaphors can play a vitally important structural role in texts rather than appearing simply as isolated expressions (see, for example, Crofts 1982 and Stienstra 1993, and also Section 3.2.3).

It is quite true that for many years the study of metaphor in translation was considered of largely peripheral interest within translation studies. For at least 15 to 20 years after the birth of the discipline as a recognisable entity around the 1960s, very little attempt was made to examine the cluster of problems raised by metaphor as an isolated phenomenon, or to discuss the place of metaphor within a broad-based account of translation in general. Furthermore, this situation could be observed across the board. Articles in scholarly publications were conspicuous largely by their absence, while few of the seminal texts on translation devoted more than mere asides to the subject. Thus Vinay and Darbelnet's *Stylistique comparée du français et de l'anglais* (1958), for example, does not discuss the matter to any great extent (see Dagut 1976:21), while Nida's *Toward a Science of Translating* (1964) devotes a mere couple of pages to the subject. Beekman and Callow's *Translating the Word of God* (1974) with its two chapters on the subject is very much the exception, and it was indeed from the middle of the 1970s that the situation began to change, albeit very slowly. Largely as a response to the single passage in Kloepfer (1967) referred to earlier (and discussed in detail in Section 3.2.2.1), in 1976 Dagut produced a major article on the subject; this was followed by a number of statements by Newmark on the matter (e.g. 1977, 1980, 1985, 1988a, 1988b). These writers' articles have proved to be highly influential in the discussion of metaphor in translation and have been cited by a large number of authors, many of whom have used them as the starting point for further discussion of the subject, either endorsing or taking issue with one or other of these earlier treatments.

In the last 30 years or so, the situation has been rather different, with questions of metaphor in translation attracting growing interest within the discipline. The articles by Dagut and Newmark were fairly quickly joined by others—most notably those of van den Broeck (1981) and Mason (1982)—while since that time an increasing number of papers have been appearing in various translation studies journals, more and more of which have started to take the interesting work carried out on metaphor in other disciplines (such as cognitive linguistics) as their starting point. This steadily increasing flow has served to keep the subject on the agenda, and in line with this development, at least another three important and widely discussed works—Gutt

(1991), Toury (1995) and Snell-Hornby (1995)—include passages discussing the area in some detail. Writers on Bible translation, it should be pointed out, have tended to be relatively prolix in their treatment of the subject. As mentioned earlier, Beekman and Callow (1974) fit into this trend, as does Larson (1984), who also devotes entire chapters to the topic.

Altogether, a dozen books that are at least substantially devoted to metaphor in translation have appeared to date: Kjär (1988), Pisarska (1989), Stienstra (1993), Kurth (1995), Samaniego Fernández (1996), Gaudio et al. (2003), Zelinsky-Wibbelt (2003), Zahid (2004), Salim-Mohammad (2007), Miller and Monti (2014), Musolff, MacArthur and Pagani (2014) and Guldin (2016). (St. André 2010 also has to be mentioned here, although this collection of articles is about metaphors for translation.) Recent years have also seen an ever-growing stream of dissertations and theses on the subject, including Zahri (1990), Fung (1994), Mandelblit (1997), Obeidat (1997), Erickson (1999), Samaniego Fernández (2000), Al-Harrasi (2001), Holm (2001), Farrell (2004), Elmallah (2008), Zhang (2009), Papadoudi (2010), Rodríguez Márquez (2010), Ureña Gómez-Moreno (2011), Najjar (2012), Omar (2012), Tcaciuc (2013), Tebbit (2013), Abu Ain (2014) and Alshunnag (2016), with a number of others currently in progress. (The fact that eight of these theses focus on Arabic-English translation is surely not fortuitous, as this is the language pair that is currently most strongly represented in the literature.)

It is therefore understandable that these days—and, of course, with the cumulative effect of the last few decades—writers on metaphor in translation no longer complain that their subject has always been somewhat neglected. As of October 2016, there were 91 items listed in the Benjamins Translation Studies Bibliography (Gambier & van Doorslaer 2016) that were published between 1968 and 2014 and contained the word *metaphor* in their titles. (There were of course many more—374, in fact—that had this term as a keyword.)

While there are a relatively large number of translation scholars with an interest in metaphor, by and large there seems to have been considerably less interest shown by metaphor scholars in translation. In terms of major works, Knowles and Moon (2006:61–72), Kövecses (2005:133ff., 2014) and Steen (2014) offer some of the few detailed discussions of metaphor in translation by metaphor scholars. In addition, there is a scattering of much briefer mentions—for example, in Lakoff (1987a:312) and Sadock (1993:52, 55). The special issue of the *Journal of Pragmatics* (issue 36:7) on 'Metaphor across Languages' is for the most part only tangentially about translation. Newmark (1985) offers one of the first discussions of metaphor in translation to have appeared in a major work devoted to metaphor rather than to translation, although his article has more recently started to be followed by others (for example, Deignan & Potter 2004; Schaeffner 2004; Samaniego Fernández 2011). To date, almost nothing of relevance to translation has appeared in *Metaphor and Symbol*.

On the worldwide web, there are relatively few significant resources, although a search carried out on 7 October 2016 using Google did reveal a huge number of references to the topic. The web also hosts a number of articles available in translation portals and online journals, such as Al-Hasnawi (2007), Brevik (2008) and Shabani (2008).

I prefer to use the formulation 'metaphor in translation' in my research because of the particular emphasis that it implies. (I also use the phrase 'metaphor translation' to talk about the activity of translating metaphor.) The more traditional way of referring to this area of study, 'the translation of metaphor', sounds prescriptive and somehow old-fashioned. 'Metaphor in translation' places the emphasis on metaphor and locates the discussion of it within translation studies where we want it—firmly in the broader context of general metaphor research. The alternative formulation 'metaphor and translation' would be another possibility and one that is more neutral in terms of the disciplines to which it is appealing. Interestingly, in the subject index to *Descriptive Translation Studies and Beyond,* Toury refers to this research topic as 'metaphor in/and translation' (1995:304).

3.2.2 *Themes in the Study of Metaphor in Translation*

Now that the discussion has achieved a fair degree of momentum, it would perhaps be sensible to attempt to summarise some of what might be seen as its main themes up to now. This will prepare the way for us to turn our thoughts to problems that require our attention. In what follows, the discussion will be organised thematically, although within each section, different authors' views are considered broadly chronologically.

3.2.2.1 *Kloepfer's 'Boldness Hypothesis' and the 'No Problem versus No Solution' Controversy*

In many ways, it was the brief passage in Kloepfer (1967) alluded to earlier that originally set the ball rolling. During a discussion of the work of a German translator of Rimbaud's *Métropolitain*, Kloepfer makes the following observation:

> Küchler [the translator] manages to preserve all the metaphors: their famous 'boldness' is no problem for the translation—on the contrary, the bolder and more creative the metaphor, the easier it is to repeat it in other languages. There is not only a 'harmony of metaphorical fields' among the various European languages, there are concrete metaphorical fields common to all mankind, but there are also definite 'structures of the imagination' on which they are based.
>
> (1967:116; translation taken from Snell-Hornby 1995:57)

Interestingly, this passage with its talk of 'metaphorical fields' and 'structures of the imagination' is in many ways parallel to some of the antecedents of

Lakoff and Johnson's theory. Indeed, as Dagut points out, Kloepfer's claim clearly reflects the notion that all human minds share the same 'perceptual and cognitive features' (1976:27). Furthermore, it possesses a certain attractiveness on a purely intuitive level. There are, no doubt, plenty of bold metaphors that can be easily translated because of the very mould-breaking nature of their boldness: what is strange or striking in one language is fairly likely to appear thus in others.[5] However, does this hypothesis constitute a universal principle of metaphor translation? And how far does it go towards providing us with a full-blown model for metaphor translation?

Kloepfer's hypothesis is repeated by Reiss (1971:62–3) and, with important qualifications, by Newmark (1985:314). On the other hand, Dagut, who uses this statement of Kloepfer's as the starting point of his first article, is in fact very scathing about it as it stands, arguing that it is 'striking in its sweeping dismissal of the problem' (1976:25), ignoring as it does the existence of the 'very numerous discrepancies of detail' between any two particular languages (1976:27). Dagut concludes that 'It is only where the linguistic and cultural components of the metaphor are shared by both SL and TL . . . that the Kloepfer-Reiss thesis can be successfully applied' (1978:103).

Possible evidence to support Dagut's contention about limiting the applicability of Kloepfer's hypothesis can be found in the example discussed by Hönig (1997a:16–17) of the phrase *ein grauer Zeitbrei* being translated in the *Financial Times* as *a grey time-porridge*. According to Hönig, the original German expression, while certainly unconventional, is nonetheless based on a combination of familiar idiomatic expressions that convey notions of monotony and lack of differentiation. The literal English translation, on the other hand, fails to convey the same overtones because not all the English lexical items lend themselves to these precise figurative extensions of meaning so that the resulting translation sounds odd and is difficult to interpret.

Misleading as the hypothesis may or may not be, it nonetheless serves to highlight a wider debate concerning the question of translatability. Following Dagut's terminology, the two approaches that this is centred on are sometimes referred to as the 'no problem' and the 'no solution' schools (1976:25). The first of these probably finds its clearest articulation in Kloepfer's statement quoted earlier. The second, with its implied inbuilt pessimism, is most closely associated with those authors—mostly coming from outside translation studies—who take a fastidious view about the possibility of translating poetry, although it also finds a degree of reflection wherever the recommendation is made that a metaphor should be translated by a non-metaphor (see Dagut 1976:25) and seems to be adhered to by Vinay and Darbelnet (1958) and Nida (1964), as pointed out by Samaniego Fernández (1996:124). Surprisingly, perhaps, in the light of his intensely polemical stance against the 'no problem' school—a position that is if anything strengthened in his later (1987) article—Dagut favours a compromise between these two positions, in which the translatability of a source-language metaphor is a function not of its 'boldness' or 'originality', but of '(1) the particular cultural experiences and semantic associations exploited by it, and (2) the extent to which these can, or cannot, be reproduced non-anomalously in TL, depending on the degree of "overlap" in each particular case' (1976:32).

Such a compromise position is also favoured by Snell-Hornby, another writer who is outspoken in her criticism of the 'no problem' school:

> . . . the answer lies somewhere between the two poles [i.e. of total translatability and total untranslatability], depending on the structure and function of the metaphor within the text. As an abstract concept, metaphor might be universal . . .; in its concrete realization, however, being closely linked with sensuous perception and culture-bound value judgements, it is undoubtedly complicated by language-specific idiosyncrasies. And therein lies its fascination for the translator.
>
> (1995:61–2)[6]

Van den Broeck also concludes that bold metaphors are more translatable than conventionalised metaphors 'to the degree that they are less culture-bound and are therefore able to dispense with *culture-specific* information' (1981:84, emphasis original). In spite of these (admittedly common-sense) contentions, Kloepfer's idea has not as yet received significant theoretical testing from the point of view of metaphor research. What factors do determine the translation procedure? It is in fact one of the aims of this research to try to find an answer to questions such as this.

More recently, Kövecses (2014) discusses a range of translation problems that derive from differences in conceptual systems, while Steen argues that any problems that arise are likely to be restricted to a specific class of metaphors, as for much of the time metaphor in translation seems to behave 'fairly civilly' (2014:13).

3.2.2.2 The Definition of 'Metaphor'

Much space has been devoted over the years to discussing what should or should not be 'counted' as a metaphor, and by and large authors have tended to focus on creative metaphors to the exclusion of lexicalised or dead ones. These latter are all those idioms, phrases and single lexical items (e.g. *new broom*, *level playing field*, *bed of roses*, *bitter pill*, and also *pupil* (= part of the eye), *tongue* (= language) and *grasp* (= understand)) an awareness of whose metaphorical origins has been lost by most speakers. Newmark is happy to take account of the entire gamut of metaphor-like phenomena and considers the problems involved in translating what he terms dead, cliché, stock, recent and original metaphors (1988b:106–12). Dagut, on the other hand, takes a narrower view, contrasting 'metaphor'—which he argues is by definition 'live' (1976:22)—with polysemes, idioms and formators (these last being items such as *despite* or *notwithstanding*, which are analysable into smaller semantic units that do not, however, give an accurate idea of the true meaning of the word). Translation of these latter three categories is, he argues, 'an area of translation that can be fully "mapped" by translation theory' (1976:24)—and is therefore of no interest to him.[7]

This is a notion that van den Broeck takes issue with, although as we shall see next, he insists that the determining factor is an expression's 'functional relevancy to the communicative situation' in which it occurs (1981:82). He coins the term 'deautomatized idioms' (1981:83) to describe lexicalised metaphors that take on a special significance in a particular context because of the clear functional role they are being assigned—as might be the case in various types of creative writing, for example.

Dagut's later (1987) article continues with the distinction between metaphor 'proper' (i.e. creative metaphor) and its 'derivatives' (polyseme, idiom, proverb: see, for example, 1987:83). In this article, he states that 'there *are* metaphors which lie beyond the limits of translation in the strict sense (excluding any form of periphrastic rendering), and that the reasons for their untranslatability may be either cultural or lexical or both' (1987:81; emphasis original). But why is this perfectly reasonable statement restricted to 'live' metaphor without any further discussion? One matter that is being investigated in this research is whether it has any validity for one of the three categories that he excludes (i.e. polysemes)—or, at least, for individual expressions within these categories. Here just as with 'metaphor proper', semantic and cultural associations can intervene, possibly in a way that will make his claim that these areas can be 'fully "mapped" by translation theory' (1976:24) somewhat wide of the mark, as stated by van den Broeck (1981:82). We are now in fact touching on issues that are going to be thoroughly discussed later on in the book. However, suffice it to say at this stage that the patterning of 'dead' metaphors into domains—and the interaction between different domains that makes all types of semantic transfer possible—is a significant area that he does not attempt to consider.

In fact, Dagut's 1987 article, which was published some seven years after the appearance of Lakoff and Johnson's *Metaphors We Live By*, betrays no familiarity with the important advances in the study of metaphor that this work ushered in. By contrast, Alexieva takes account of research in cognitive linguistics carried out in the 1980s and 1990s to propose a nuanced and well thought-out domain-based approach:

> . . . we find differences across languages and cultures, firstly, in the internal structuring of the various domains of knowledge and experience; secondly, in the interconnectedness between domains and the way one domain is mapped onto the other; and thirdly, in the stability of both the inner structuring of the domains and their interconnectedness.
>
> (1997:141, emphasis removed)

Although she is ostensibly writing about word play in this article, what she says can be applied equally well to metaphor—both the original and lexicalised varieties. For her, the asymmetry that she discusses is both interlingual and intercultural in nature, and takes in not only the creative, original linguistic units that she is specifically writing about but also a range of more commonplace types of linguistic usage. In terms of the translation solutions that she suggests, therefore, what becomes important is to consider meanings of words or phrases not in isolation but in the context of their 'cognitive and experiential implications' (1997:142). The idea of transfer, interaction or confrontation between the two different domains—an interrelation that somehow needs to inform the solution that appears in the translated text—becomes paramount, replacing the notion that a piece of figurative language is the one-off isolated result of creative inspiration on the part of the writer or speaker. This broader view of metaphor is generally taken by scholars whose approach to metaphor in translation is generally in line with that of cognitive linguistics (see, for example, Mandelblit 1995; Samaniego Fernández 1996; Al-Harrasi 2001; Farrell 2004; Schaeffner 2004; Al-Hasnawi 2007; Tobias 2009; Ding, Noël and Wolf 2010; Papadoudi 2010; Shuttleworth 2011, 2013).

3.2.2.3 The Single Metaphor versus the Metaphor Network: The Nature of Metaphor within Text

A clearly related matter is the question of whether translation scholars have viewed the problems of metaphor translation as revolving round finding ways of translating isolated one-off metaphors or have been more alert to instances where one single structurally significant metaphorical idea underlies a whole series of metaphorical expressions, thus becoming deeply woven into the fabric of the text in which it features.

It has to be said that in the past many writers have fallen into the former camp. Neither Newmark nor Dagut, for example, appear to be concerned with those instances where whole patterns of metaphors seem to be sanctioned by one-and-the-same underlying mapping or where a single metaphor may pervade an entire text. Even for Snell-Hornby—who talks about the 'structure and function' of metaphors within texts (1995:61)—this is a facet of metaphor in translation that does not feature particularly prominently. That is not to say that many of these writers have not been interested in the textual relationships that (admittedly individual and free-standing) metaphors enter into, but simply that the focus of their investigations has not been on the way a single mapping can structure an extended passage or even an entire text.

One of the first writers to emphasise the fact that metaphors can play a vitally important structural role in texts was Crofts who, with specific reference to Bible translation, suggests the five-way classification of incidental, repeated, extended, semantic and symbolic metaphors (1982:16). Of interest to us here are the middle three, which between them cover a range of closely related types of patterning—even though the distinction between some of these categories is not sufficiently developed in this very brief article.

However, it was another writer on Bible translation, Stienstra (1993), who was one of the first to argue forcefully that the view that saw metaphors occurring in texts solely as one-off isolated phenomena was inadequate. Stienstra observes that 'no attention has been paid so far to the importance of preserving all the various manifestations of a metaphorical concept in a particular text' (1993:216); she argues that since the units studied have been 'simply inadequate' it has inevitably led to generalisations of a 'spurious' nature being arrived at (1993:217). Her recommendations for improving this unsatisfactory situation are as follows:

> If we want to discuss the translatability of metaphor at a more advanced level than that of the individual example, we will have to show ourselves aware of the fact that many, if not most, interesting metaphors form networks which are both systematic and dynamic. It is only by investigating whether such a network can be transferred to another language, that we can make progress in answering the question as to the translatability of metaphor.
>
> (1993:215)

Whether or not the ultimate purpose of investigating metaphor in translation is to look at the question of translatability, in the light of all the interesting work that has been produced within the cognitive branch of metaphor research within the last 20 years or so, it would be somewhat blinkered not to take this recommendation with the utmost seriousness: if certain metaphors are repeated throughout a text in a structurally significant manner—as is manifestly the case in some texts—surely it would make sense to take this fact into account when studying the phenomena related to metaphor in translation?

Slightly more recently, Hatim and Mason (1997) present an example of an extended metaphor in the form of a page-long extract from a message by Ayatollah Khomeini, in which the late Iranian cleric makes repeated use of the imagery of blood in order to exhort the staff and students of religious seminaries to higher levels of devotion, militancy and vigilance (1997:148–50). Presented as what Venuti (2008) would term a 'foreignising' translation, the English version basically preserves each source-text instance of the blood metaphor. Hatim and Mason focus on the fact that this ensures that an important cohesive chain is maintained in the target text (1997:151); however, for my purposes, it is also of interest that in this instance the translator elected to retain certain striking metaphorical expressions (such as 'soaked in blood on the pavements of bloody events'), which in the target text have become divorced from a particular mapping that lent them special significance for the text's source-language readers.

3.2.2.4 Metaphor and Text-Type

Another major issue is that of the text-types in which metaphor is supposed to play an important role. Hardly surprisingly, much of the interest in metaphor in translation has been concentrated on literary texts, with many scholars—such as Dagut (1976), Mason (1982), Alvarez (1993), Kruger (1993), Fung (1995), Kurth (1995), Monti (2006), Fouces González (2007), Jay-Rayon (2007), Maalej (2008), Tobias (2009) and Omar (2012)—focusing on this particular area. In line with the more traditional view of literary metaphor, some but by no means all of these writers have tended to concentrate on original one-off metaphors. After these literarily oriented approaches, the second major focus of interest has been religious texts. A large number of Bible translation scholars have written extensively on the subject, including, for example, Beekman and Callow (1974), Crofts (1982, 1988), Gutt (1991), Larson (1984) and Stienstra (1993). Zahri (1990) and Najjar (2012) have conducted major studies of metaphor within the context of translating the Quran, and other scholars such as Faiq (1998) and El-Zeiny (2011) have also written on this topic.

If one is to mention a third area that has attracted the interest of scholars, then it is probably that of politics, on which both Schaeffner (1996, 2004) and Al-Harrasi (2001) have focused their attention—Schaeffner with discussions of the translation of EU documents and Al-Harrasi with an analysis of official Omani translations of speeches by the Omani head of state Sultan Qaboos bin Said and of translations of speeches by Saddam Hussein carried out by the translation service of the CIA. As for other text-types and subject areas, van Vaerenbergh (1996) and Tcaciuc (2013) consider economics and banking texts, Pisarska (1989) covers a wide range of popular scientific and other non-literary texts, while van den Broeck (1981) devotes a fair amount of space to a discussion of the relative translatability of metaphor in literary and scientific texts, concluding that metaphorical expressions in the latter type of text do not usually constitute much of a problem for the translator. Metaphor translation in scientific discourse has been tackled by Stambuk (1998), Temmerman (2002), Vandaele (2002), Scarpa (2004), Faber and Márquez Linares (2005), Papadoudi (2010), Shuttleworth (2011) and Vandaele and Béland (2012). However, the relative lack of interest in non-literary writing that we see in most authors clearly goes hand in hand with the emphasis that is frequently laid on 'creative' metaphor, and there seems to be a widely held view that the translation of metaphor in technical or scientific texts does not constitute a problem. Van den Broeck (1981) at first glance appears to adhere wholeheartedly to this opinion. Distinguishing between 'creative' and 'non-creative' metaphor, he states flatly that 'in the former metaphor *as such* is functionally relevant, whereas in the latter it is most likely not to be' (1981:78, emphasis original). He then goes on to argue,

> The extreme positions are occupied by literary language on the one hand, and the language of science on the other hand. . . . In scientific

> discourse bold metaphors are very unlikely to occur. Lexicalized ones will of course be unavoidable, but it goes without saying that for a translator there is no problem in rendering them (metaphorically or non-metaphorically).
>
> (Van den Broeck 1981)

As it stands, this contention—which is diametrically opposed to that of Kloepfer—does not seem to give serious consideration to the possibility that the translation of metaphor in scientific texts might constitute a problem. Later on in his article, however, van den Broeck does go some way towards moderating this somewhat categorical statement. Building on the important notion of functional relevance—which he argues is not necessarily synonymous with 'boldness' (1981:76)—he states that the treatment of lexicalized metaphors 'will entirely depend on their functional relevancy to the communicative situation in which they occur' (1981:82). Thus, for example, in certain contexts (this happens most typically in poetry) an idiom such as *barking up the wrong tree* can assume great functional relevance by having both its literal and its more usual, figurative meaning simultaneously activated.

However, van den Broeck still considers the translation of items of this type when they occur in 'non-creative language' simply as a matter of consulting a suitable dictionary for 'either a corresponding polyseme or idiom, or an equivalent non-metaphorical expression' (1981:82). Pisarska (1989) takes rather a different view of the question of metaphor in technical texts. Rejecting most of the categories into which metaphors have been variously divided, she suggests that both original and lexicalised types 'may pose similar problems in translation' (1989:57). She also observes that the presence of metaphorical expressions in non-literary texts is 'undeniable' and argues that, if only for the sake of obtaining a complete picture, the translation of conventional metaphor in such texts deserves to be studied (1989:1).[8]

In the case of most non-literary text-types (including most popular scientific texts—which have a reasonable chance of including fairly well-developed, striking metaphors) we might perhaps anticipate that a fairly 'literal' translation strategy that preserves the metaphor is the norm, or indeed the default position. Indeed, writing on the basis of her analysis of the books in her corpus, Pisarska states that 'all examples of metaphorical uses found in the examined texts are those of lexicalized metaphors which have lost their unique, poetic character and have become part of the established semantic stock of the language' (1989:68), and concludes that the degree of translational creativity here is low (1989:98). Consequently, any occasion on which the translator feels unable to render a metaphorical network in broadly 'equivalent' terms should be deserving of our very serious consideration.

In spite of the conclusions that Pisarska draws on the basis of her study, the present work takes as its starting point the conviction that metaphor plays at least a potentially significant role within popular scientific discourse, given the way in which metaphorical conceptualisations can model or even channel thought in a sometimes covert manner (and to an as yet unquantified extent). Probing the effect that translation can have on significant networks of conceptual metaphors within texts is of course one of the aims of my research.

3.2.2.5 Translation Procedures

Van Doorslaer (2007:226–7; see also Munday 2012:22–4) typifies the manner in which the terms 'strategy' and 'procedure' are used in contemporary translation studies by indicating that the former refers to the general approach to translating a text that is adopted by a translator (e.g. literal translation, idiomatic translation), while the latter is reserved for a means of rendering a specific textual feature or a solution to a particular problem (e.g. compensation, implicitation, etc.). It is with translation procedures—which bring about the precise shifts that metaphorical expressions undergo during the translation process—that the research concerns itself. As stated above, a translation 'effect', a third but less frequently used concept, is simply a change in the text—whether global or localised—that has been brought about during the translation process. Finally, Pym (2016) chooses the word 'solutions' as an alternative to 'procedures'.

It would of course be possible to produce a highly detailed taxonomy, reflecting, for example, most or all of the many shifts identified by van Leuven-Zwart, including shifts between the general and the specific, the abstract and the concrete and the objective and the subjective, as well, for example, as 'explicitation, implicitation, amplification, reduction, addition, deletion, intensification and archaization' (1989:153). Alternatively, the lists proposed by van Doorslaer could also provide the starting-point for such a taxonomy. Each of these would, potentially, take the research off in a different direction.

In his detailed discussion of 'solutions', Pym considers theoretical writings in a wide range of languages and time periods. What he produces is a set of solution types that are expressed in simple terms and placed at a high level of generality. His types, which have been derived for the most part inductively, describe how translators transform a text and are applied to situations where a choice has to be made (2016:232–3). They are optional in nature (so Catford-like shifts do not count), applicable to many language pairs and—music to the ears of any researcher—can be broken down into an unlimited number of subcategories (2016:219).

The list of solution types that he arrives at is as follows:

1. Copying words (e.g. copying sounds, morphology, script)
2. Copying structure (e.g. copying prosodic features, fixed phrases, text structure)
3. Perspective change (e.g. changing sentence focus, semantic focus, voice)
4. Density change (e.g. generalisation, specification, explicitation, implicitation)
5. Compensation (e.g. new place in text)

6. Cultural correspondence (e.g. corresponding idioms, culture-specific items)
7. Text tailoring (e.g. censorship, addition, omission)

(2016:220)

This is an interesting list, and one that contrasts somewhat with another recent scheme, Pedersen's proposed six-procedure 'baseline' of retention, specification, direct translation, generalisation, substitution and omission (2011:73–4). While Pedersen focuses on single culture-bound textual items, Pym also takes in phenomena that involve larger chunks of text or relate to the macro-level. In addition, some of Pym's categories are superordinate to some of Pedersen's (e.g. density change vs. generalisation and specification) and vice versa (e.g. retention vs. copying words and copying structure)—the kind of incompatibility that you almost inevitably encounter when you compare two systems of categories.

On the question of procedures for translating metaphor, Pisarska discusses the various ways in which translators handle metaphorical language in a range of different contexts. Talking specifically about the type of conventional metaphor that one typically finds in non-literary texts, she states,

> The numerous possibilities of rendering these seemingly non-problematical expressions are signalled by the number and variety of procedures translators resort to in the renditions and their value for the idiosyncratic style of writing of a particular author is confirmed by the fact that in the vastly prevailing number of cases they are retained in the translations.
>
> (1989:1)

Pisarska also describes an '*ad hoc* survey' (1989:63–4; emphasis original) in which 15 English expressions drawn from Lakoff and Johnson (1980:7–8) that illustrate the metaphor TIME IS MONEY were translated into seven languages (three of which were not Indo-European). Interestingly, it reveals a remarkably high level of similarity between the various languages. Equally interesting, however, is the fact that one expression, *Is that worth your while?*, is not rendered directly in any of the seven languages, suggesting that even if there is general cross-language agreement as to the acceptability of a conceptual metaphor, there are always likely to be differences of detail from one language to another. As far as I am aware, this is a 'one-off' experiment that does not form part of a detailed large-scale comparative investigation, but the results are nonetheless thought provoking. Another experiment from the 1980s (Snell-Hornby 1995:59–60), conducted this time with the researcher's own students, focuses on the simile chosen by groups of German and English speakers to describe the image metaphor set out in Figure 3.1, which first appeared in an article in the *Neue Zürcher Zeitung* (7 April 1978), where it was intended to depict how Catholic and Protestant communities in Belfast are arrayed round the city centre and separated from each other by the city's main streets:

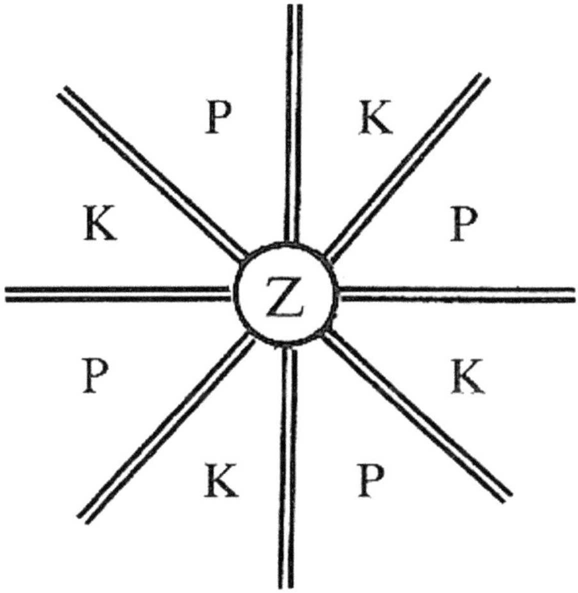

Figure 3.1 The Arrangement of Catholic and Protestant Communities around the centre of Belfast. Taken from Snell-Hornby (1995:59) (K = Catholic, P = Protestant, Z = centre)[9]

The original German sentence was supplied to participants with the word used to describe the arrangements of the streets removed. According to Snell-Hornby, both groups answered 'without hesitation, and almost unanimously' (1995:60), with the German speakers correctly supplying the original word *strahlenförmig* ('in the shape of rays') and the English speakers seeing the diagram as resembling the spokes of a wheel. In spite of the English group's near-unanimous opinion as to what felt 'right' in this context, nearly all of the over 150 students who translated this text opted for the literal 'like rays of the sun' or 'like rays of a star' as the translation for this word (ibid.), thus indicating the inherent tension that exists between what native speakers consider to be the correct situational equivalent in their language and what often ends up appearing in a translation because of the influence of the source text (1995).

Much effort has also been devoted over the years to constructing classifications to account exhaustively for the procedures that translators employ when translating metaphorical expressions. Large amounts of space are not available to discuss this area, and so one example will probably suffice. Newmark's classification of the procedures that translators use to translate what he terms *stock* metaphors is very representative of the type of scheme that has been proposed. In all he identifies eight procedures, which he lists in order of preference:

1. Reproducing the same image in TL, 'provided the image has comparable frequency and currency in the appropriate register'
2. Replacing the image in SL 'with a standard TL image which does not clash with the TL culture'
3. Translating the metaphor by a simile, 'retaining the image'
4. 'Translation of metaphor (or simile) by simile plus sense (Mozart method)' (i.e. the addition of explanatory material: for Mozart, a piano concerto had to please 'both the connoisseur and the less learned')
5. Conversion of the metaphor to sense
6. Modification of the metaphor
7. Deletion
8. Using the same metaphor combined with the sense

(1985:304–11)

While this list clearly has value as a set of guidelines for translators, what it tries to achieve does not coincide with the aims of descriptive translation studies as it is essentially prescriptive in nature. Similar schemes have been proposed by a range of authors including Beekman and Callow (1974:145–9), Larson (1984:254), Pisarska (1989:76), Alvarez (1993:484–7), Zhang (2009:105, 107), Papadoudi (2010:279–81) and Tebbit (2013:63–5). However, as a result of the considerable work already devoted to producing classifications of this type, this area is probably quite well charted: while Newmark's list does have a few gaps (such as the theoretical possibility of translating a metaphor by a metaphor plus sense), any minor modifications that have been made are little more than footnotes to his scheme. The one significant exception seems to be Toury's 'prospective' view of seeing metaphor as a solution rather than a problem, which produces two more possible scenarios: translating a non-metaphor by a metaphor and adding a metaphor in the target text when there is no 'linguistic motivation' in the source text (1995:83). With these additions, though, it would seem that this approach has been taken about as far as it can be in terms of the major procedures that such lists attempt to classify—although they can never be exhaustive, as it is always possible to find further one-off special situations that have not yet been accounted for.

Another point about Newmark's list is that, while all the procedures are theoretically possible, some in fact rarely occur, which means that it is perhaps difficult to use it as a tool for descriptive research in an unmodified form.

The other type of procedure list that should be mentioned is more descriptive in nature and is firmly grounded in metaphor theory as well as translation research. One of the earliest examples was produced by Mandelblit (1995), who observes that target-language versions of metaphorical expressions can conform to either the 'Similar Mapping Condition', in which case source and target use 'the same general metaphorical mapping' or to the 'Different Mapping Condition', where different ones are used (1995:491).

Along similar lines, another list of procedures that is worth examining is that of Al-Harrasi (2001:277–88). This list is radically different from all its predecessors. What is new about his taxonomy is that it is deeply informed by conceptual metaphor theory, hinging more specifically on the interplay between image schemas and rich images:

1. Instantiating the Same Conceptual Metaphor
   - 1.1 Same Image Schematic Representation
   - 1.2 Concretising an Image Schematic Metaphor
   - 1.3 Instantiating in the TT only a Functional Aspect of the Image Schema
   - 1.4 Same Image Schema and Rich Image Domains
   - 1.5 Same Rich Image Metaphor but Alerting the Reader to the Mapping
   - 1.6 Using a Different Rich Image that Realises the Same Image Schema Realised by the Rich Image in the Source Text
   - 1.7 From the Rich Image Metaphor to Image Schematic Representation
   - 1.8 Same Mapping but a Different Perspective
2. Adding a New Instantiation in the Target Text
3. Using a Different Conceptual Metaphor
4. Deletion of the Expression of the Metaphor

(2001:277–88)

As far as I am aware, this is one of the most detailed attempts to date to produce an alternative taxonomy based on metaphor theory, and one that opens up a whole new possible direction for research. Interestingly, Al-Harrasi also talks about 'procedures', even though what he lists are probably not actions that are consciously implemented by translators. His systematic pairing of image schemas and rich images has an intuitive appeal in many respects, although one possible criticism relates to its tendency to overgeneralise the concept of the image schema (or maybe it is simply that cases that do involve an image-schematic metaphor have been specified in much greater detail than those that do not). Image schemas do not underlie all conceptual metaphors, as Al-Harrasi indeed accepts (see 2001:277), and as I hope to demonstrate below; indeed, it is likely that a large proportion of metaphorical expressions involve some kind of rich image. In addition, even when an expression in a text is ultimately derived from an image schema, there is often a complex 'network of intermediate mappings' that intervenes (Evans & Green 2006:301).

Al-Harrasi's scheme has now been joined by other noteworthy examples, such as those by Schaeffner (2004:1267) and Papadoudi (2010:279). All of these works demonstrate a clear awareness of the distinction between micro- and macro-levels. Most of them also define a set of theoretical parameters to which they pay particularly close attention in their analysis of metaphor in translation and on the basis of which they propose a list of translation procedures. Along similar lines, the present work will also investigate the interplay between parameters and procedures.

3.2.2.6 Issues of Culture

The study of culture is of no greater or less intrinsic interest or complexity than that of language. Furthermore, how the two concepts interact and intertwine with each other is a matter of great intricacy and has yet to be fully understood (although it is being probed in detail by writers such as Wierzbicka 1992, 1997; Goddard 2006; Goddard & Wierzbicka 2014). In the literature, questions that properly fall within the domain of language are sometimes incorrectly attributed to cultural factors (as argued, for example, by Dobrovol'skij & Piirainen 2005:210) and presumably also vice versa. As has already been made clear, the emphasis of this research is on language, although questions relating to culture are also addressed whenever they present themselves.

The concept of culture and its influence on translation is very broad. Katan, for example, sees it as a wide-ranging notion that affects the translation process on many different levels. Cultural background can determine a person's basic communicative stance (e.g. use of understatement, directness of expression; 1999:210–41). The appropriateness of different levels of formality and distance can also vary from culture to culture (1999:-177–209), as can the extent of the use of nominalisation as a basic means of organising information in sentences, for example (see 1999:113–4; see also Bennett 2011). Individual items of lexis can possess significant cultural dimensions (1999:151–2). (This is a matter that Wierzbicka 1997 examines in great detail.) In addition, there are six hierarchically arranged 'logical levels' within which all biological and social systems can function (Katan 1999:37); these are environment, behaviour, capabilities, values, beliefs and identity (1999:45–61). Finally, as stated in the previous paragraph, there is the whole question of how language, thought, culture and reality relate to each other (1999:74–5). This raises important questions of linguistic universalism or relativism, at one end of which we see views championed by scholars such as Pinker, who vigorously denies the possibility of any link between the language that a person speaks and the way in which he or she views the world (see, for example, 1994:58), and at the other the so-called strong version of the Sapir/Whorf hypothesis, according to which we are completely at the mercy of the view of reality that is imposed on us by our native language (see, for example, Whorf 1956:213). The very existence of translation as a real-world activity is of course a significant argument against both such extremes.

Katan does not consider the question of metaphor in detail. However, two specific points need to be made. The first is that the question of interlingual metaphor variation needs to be addressed in the light of the relatively universalist view regarding how metaphor exists in different languages that has been adopted by many metaphor theorists. To counter this, Gibbs presents metaphor as cognitive networks that 'extend beyond individual minds and are spread out into the cultural world' (1999b:146). According to him, 'cognition arises, and is continually re-experienced, when the body interacts with the cultural world' (1999b:162). Gibbs concludes by acknowledging 'the culturally embodied nature of what is cognitive' and suggests that 'there is much less of a difference between what is cognitive and what is cultural than perhaps many of us have been traditionally led to believe' (1999b:162).

Along similar lines Kövecses (2005) sees metaphor as existing in culture at the interface between universality and variation: 'universality' because of the cognitive source of much metaphor (2005:17–64) and 'variation' because of the alternative metaphors that exist in different cultures (2005:67–87) and the variability of metaphor within individual cultures—for example, for diachronic, social or stylistic reasons (2005:88–113). Talking about figurative language in general, Dobrovol'skij and Piirainen also arrive at a very similar conclusion:

> . . . the real behaviour of a given idiom is not governed by knowledge of a general kind (e.g. by knowledge about containers), but by more concrete, culture-specific knowledge. . . . At the level of the CONTAINER metaphor, all expressions for ANGER are identical. However, they differ from each other in their linguistic behaviour: they may show different restrictions of use in one language, or, from a cross-linguistic viewpoint, they may not be considered full equivalents.
>
> (2005:130)

In fact, in terms of sheer numbers, a view of metaphorical language that accommodates cultural variation now appears to be the predominant one, even if some of the most dominant voices still hold a much more universalist position.

The second point that needs to be made with reference to metaphor in culture regards the cultural turn in translation studies. In line with much of the other work done on metaphor in translation, this study is more deeply influenced by metaphor research than by most of the new 'culture-oriented' approaches to translation listed in Section 3.2.1.1. However, in spite of its largely linguistic orientation, this research does not aim to turn the clock back to the early 1980s. Where necessary, the target context is taken into consideration and translation variants not opted for are given due consideration. In addition, where appropriate, conclusions that go beyond the purely linguistic are drawn (see, for example, Interlude One).

This section has served partly as a kind of justification of the project in the light of some of the more 'universalist' metaphor theory on the one hand, and in that of the cultural turn on the other. In view of all that has been discussed, we can delineate the corpus of metaphorical expressions that this work analyses as being determined by a combination of relatively universal cognitive factors, the 'accumulated cultural experience of the members of [a] language-community' and the '"institutionalized" semantic associations of the items in their lexicon' (Dagut 1976:32). In other words, the influence of culture is deemed to be quite considerable in this area.

3.2.2.7 The Question of Forming a Separate 'Theory of Metaphor Translation'

A number of writers have maintained that it would be desirable to treat metaphor in translation as a distinct theory. Thus Dagut, for example, argues that 'a theory of translation (like any linguistic theory) compares language *systems* (on the level of "competence"), not individual instances of "performance"' (1987:82, emphasis original). I would argue, however, that this is a very peculiar statement to make, given the preference of the discipline for matters of *parole* (broadly equivalent to the 'performance' in Dagut's argument) and norm-based behaviour rather than *langue* (corresponding to 'competence') that was identified earlier. Dagut goes on to state, 'It is precisely the unique system-violating character of metaphor that sets it apart from other phenomena of language and therefore requires a special theory to account for its translation' (1987:82), although what the basis for this special pleading actually amounts to is not entirely clear.

Pisarska is also supportive of the idea of a separate theory for metaphor translation, although she concedes that theorists are not currently in a position to formulate one (1989:115–6)—or at least not at the time that she was writing. Al-Harrasi is not only a proponent of a separate theory, but even of a distinct discipline (2001:264–9), which he proposes naming 'MiT [i.e. Metaphor in Translation] Studies' (2001:267).

On the other hand, Mason—against whom all of these three scholars react—states, 'There cannot be a theory of the translation of metaphor; there can only be a theory of translation, and that theory has to allow room for the notion of the purpose of translating each new text' (1982:149). The question will be returned to in the course of the discussion, but at this stage, it appears that metaphor in translation is simply too specific for one to make it the object of such an initiative, and doing so would go against the general trend towards integration and the formation of an interdiscipline that can be observed over the last 20 years or more within the discipline (see Snell-Hornby, Pöchhacker & Kaindl 1994; Snell-Hornby 1995, 2006). Moreover, metaphor is in many ways comparable to a range of other phenomena, such as word play, allusion or cultural items, and I believe it would lead to too great an atomisation of the discipline if each were to become the subject of its own separate theory. On the other hand, in the terms of Holmes's map that was discussed earlier in the chapter, metaphor in translation clearly forms the subject of a partial theory, although in this case its *raison d'être* would be to feed into the larger discipline.

3.3 A Short Discussion of Methodology

> Very often, target text metaphors were assessed 'in terms of some *preconceived* balance between the features of the original metaphor, mainly meaning, constituents and [type and extent of] metaphoricity. If the behaviour of a body of metaphors under translation was studied at all, the material thus collected was normally approached as if it constituted a mere reservoir of isolated "examples" rather than an organized whole, testifying to more or less regulated behaviour under certain circumstances.'
>
> (Toury 1995:81, emphasis and interpolation original)

This quotation from Toury indicates his preferences with regard to descriptive translation studies research, whether or not it is to do with metaphor in translation: for him, what is important in descriptive translation studies is its focus on an open-minded but contextualised analysis of 'behaviours', rather than a collection of observations about individual examples (even if the former often proceeds from the latter). This basic approach is also of fundamental importance for my research, in which the notion of systematicity—in the sense of generalisations that can predict what is likely to occur in translation—is of great interest.

In this section, I will look briefly at various aspects of the methodology I have used in this research, leading up to an introduction to the multilingual approach to translation research that I have developed.

3.3.1 Text Selection

What is immediately striking about many (but not all) *Scientific American* articles is the high level of metaphoricity that they possess. However, when I was planning the composition of my corpus, not only did it seem highly unlikely that every article would contain similar numbers of metaphorical expressions, but I was also doubtful as to whether each subject area would be characterised by the same levels of metaphor use. Relying on my intuition, in my selection of texts for inclusion in the corpus, I considered subject areas such as neurobiology, biotechnology, cosmology, artificial intelligence and information technology to be potentially of interest, but fields such as pharmaceuticals, therapeutics, gadgets and armaments less so. Whether or not I was right to exclude these topics, those that were included yielded a high proportion of articles that contained large numbers of metaphorical expressions.

Later in the process, articles were informally ranked on the basis of their metaphoricity and some of the more metaphor-rich were analysed in detail. My eventual selection of texts reflects my earlier intuition, and, whether or not they are in fact the most metaphor-rich, my general observation is that it was possible to extract an impressive collection of metaphorical expressions from a relatively small number of them.

3.3.2 Data Collection

As stated in the introduction, my data comprises six *Scientific American* articles that appeared between January and May 2004, along with their translations into the six target languages included in the project. Altogether, it totals around 24,000 words of English text, with corresponding (or sometimes slightly smaller) totals in each of the other languages. The reason why the word count in some target languages is lower is that not all articles are translated into all languages, as there is no French version of 'Synthetic Life' and no Italian one of 'The Addicted Brain' or 'Freud Returns'. This has been taken into consideration on the rare occasions that it has proved necessary.

The metaphorical expressions that I found in the six English articles were entered exhaustively into my data spreadsheet and then the corresponding text segments from the translations were added. Furthermore, as major metaphorical structures gradually began to emerge in the course of the identification process (such as those associated with nature), the WordSmith Tools concordancing software was used to search the entire corpus for further instances of relevant keywords derived from examples that I had already located. This approach was experimental and was not implemented on any large scale, but it was the stage that permitted me to identify metaphorical expressions that had been added in translation.

English metaphorical expressions were identified from the six articles in the quantities set out in Table 3.1:

Table 3.1 The Contribution and Subject Area of Each Article

Article	*Number of English metaphorical expressions*	Scientific American *subject category*
'The Other Half of the Brain'	107	Neurobiology
'Evolution Encoded'	63	Biotechnology
'The Addicted Brain'	37	Neurobiology
'Synthetic Life'	36	Biotechnology
'Freud Returns'	26	Neurobiology
'Decoding Schizophrenia'	10	Neurobiology

Among the specific target domains represented by the English metaphorical expressions, 73 concern neurons, 68 genes, 18 nature and 14 the nervous system, while 30 can be termed 'miscellaneous'—a category that includes some 'general-purpose' scientific metaphorical expressions (e.g. *intellectual framework*, *think outside the box*, etc.) and some that are not specific to science (e.g. *trigger*, *underpin*, etc.).

Each of the articles is written by a different author (or pair of authors)—all of whom are research scientists except for Wayt Gibbs, the author of

'Synthetic Life', who is a professional science writer. One of the main advantages of focusing in detail on a small number of articles is that some mappings are represented by large numbers of expressions, and in this way it is possible to be sure of capturing them all.

Candidate expressions that were accepted as metaphorical were added to an Excel spreadsheet as multilingual examples along with detailed meta-information and classified according to translation procedure and the six theoretical parameters; the same happened with the relatively small number of expressions that were found to have been added in translation. Multiple instances of one-word metaphorical expressions were not listed exhaustively, although repetitions were sometimes recorded to reflect the fact that an expression occurred particularly frequently. For reasons discussed in Section 1.2.3, most examples of dead metaphor were not included in the spreadsheet. In all, the database of examples comprises a total of 279 rows (with one multilingual example occupying each row). Of these, 268 contain an English metaphorical expression along with between four and six translations. The other 11 rows consist of a non-metaphorical English phrase that is translated by a metaphor in one (or occasionally more) target languages. In all, given these two facts, this data gives rise to a total of 1,505 target-language examples (these include instances of removal or omission). With a total of 57 columns defined, the dataset consists of 15,903 data points (although up to 10% of this total are empty for various reasons).

3.3.3 Metaphor Identification

> Find all the metaphors in a text: This computer is only a piece of plastic and this software can't do that. It'd be nice if it could, though.
>
> (WordSmith Tools 3.0 Help topic)

> Spotting the presence of metaphor is often quite intuitive but producing an infallible test for metaphor is practically impossible.
>
> (Stallman 1999:34–5)

Some progress has in fact been made with automatic and semi-automatic metaphor identification (see Fass 1991, Mason 2004, Berber Sardinha 2006, and Ureña Gómez-Moreno & Faber 2011, for example), although no system is as yet capable of producing a reliable list of candidate metaphors unaided from any discourse type. For this reason, metaphorical expressions have been identified manually in this project. Intuition is indeed involved in this process, and yet it needs to be supported by a set of procedures that are as rigorous as possible, even if infallibility is well-nigh impossible.

The homing in on, gathering and classification of data is likely to be a complex, multi-stage process, with numerous changes of direction dictated by the need to be selective from the sheer volume of information and one's own developing understanding of how to create appropriate methodologies for the study of metaphor systems.

The procedure for classifying all candidate expressions as metaphorical or otherwise is the same regardless of how they were identified and is fraught with difficulties. In many cases, it is no easy matter to decide whether or not a linguistic expression should be considered to possess a metaphorical meaning. For instance, is a *population* of glial cells being referred to in terms of a human 'population' through the use of this word? When cocaine *targets* the brain's reward system, are we witnessing a (very slight) figurative extension of meaning? The corpus contains a scattering of such marginal instances.

Although preceding it by a couple of years, the process of metaphor identification that took place was broadly in line with the 'Metaphor Identification Procedure' (MIP) proposed by the Pragglejaz Group (2007:3). The approach was inductive, in that all mappings and other high-level structures were posited on the basis of the metaphorical expressions that were identified rather than trying to fit those metaphorical expressions into a pre-determined framework of categories (Pragglejaz Group 2007:33–4). In essence, according to the Pragglejaz approach to metaphor identification, if a lexical unit's usage contrasts with a 'more basic contemporary meaning'—one that is more concrete, related to bodily action, more precise or historically older, even if it is not the lexical unit's most frequent meaning (Pragglejaz Group 2007:3)—then the expression is considered metaphorical. In 2010, MIP was superseded by an updated procedure known as MIPVU (or MIP Vrije Universiteit: see Steen et al. 2010).

In the case of the present project, English candidate metaphorical expressions were initially identified by two people, with an individual wording only being included in the master list if agreement was reached. On the other hand, decisions regarding the expressions in the various target languages were generally taken by me alone, but with suitable reference material and/or native-speaker informants being consulted where the need arose.

3.3.4 Categorising and Analysing the Data

Once the data had been assembled and decisions taken about metaphoricity, a multiple classification of each metaphorical expression—in terms of each metaphor parameter and according to translation procedure—needed to take place. Needless to say, membership of a category was not always absolute, as there were often degrees of belonging (this was particularly the case with the richness parameter).

In terms of the translation procedure, I initially marked all instances in which I judged that an expression had been modified in some fairly significant manner, although in the event this has only been used at one particular point in the discussion. For counting purposes, these examples were recategorised as having been 'retained'. Together with removal, omission and addition, these general categories have been used in order to obtain an idea of broad translation trends across the data, which are as set out in Table 3.2:

Table 3.2 Distribution of Broad Translation Approaches

	Total number of examples (all six target languages taken together)
Retained	908
Removed	293
Omitted	290
Added	14
total	1,505

If a metaphorical expression is removed, this means that it has been replaced by identifiable non-metaphorical textual material. On the other hand, omission is defined as occurring when a metaphorical expression is totally missing from the target text, or, in other words, when no identifiable trace of it remains in the translation. Table 3.2 demonstrates how common these two approaches are, as together they account for nearly 40% of all translation examples. The approach of addition is almost certainly under-reported because of the way in which examples were collected.

In line with Olohan's recommendation (2004:86), a combination of qualitative and quantitative approaches are used, although with the emphasis being squarely on the former. My analysis focuses on both single examples and clusters of related ones as appropriate. Although in some respects the latter is preferable, the number of suitable examples is limited and so as not to be constantly recycling the same ones, I have decided to take this two-pronged approach. Similarly, I sometimes looked at all relevant examples, while on other occasions I only examined samples of data. For checking purposes, I have used Google Advanced Search on a few occasions. If performed with care, this use of the web as a corpus surrogate (Bernardini, Baroni & Evert 2006:10–11) can yield valuable insights; furthermore, although not perhaps ideal, it has been a practical necessity given the time that would have been needed to assemble seven separate language-specific reference corpora.

Finally, our knowledge of the possible overlap between the roles of the translators and the editors in the different editions of *Scientific American* is somewhat patchy. This, combined with the lack of information on who the translators are in some of the editions, has done much to determine the kind of analysis that is possible and may even have played a role in shaping the multilingual approach that is being used.

3.3.5 The Multilingual Approach

As the research progressed, the emphasis gradually shifted from observing the characteristics of translation into each of the target languages (where it had only partially been in any case) to looking almost exclusively at cross-lingual commonalities. The use of a multilingual approach—by which is meant analysing parallel translations of a single source text or group of source texts into a range of different languages that are being studied alongside each other—in order to increase both the number and generality of examples available for analysis is an important aspect of my research (and one that is not only of relevance to metaphor in translation, of course).

The 'gearing' effect that this approach offers means that a small number of source-language expressions give rise to many such expressions in the target language and provide access to the parallel work of a multiplicity of translators without attaching undue significance to the habits and preferences of any one individual. In addition, the major advantage that this approach gives of permitting the researcher to focus on less language- and culture-specific aspects of translator behaviour, or, in other words, on metaphor translation tendencies that may transcend the constraints of a single translation context (to the extent to which that is possible), will hopefully offer us a useful means of probing the scope and limitations of translators' interventions. The idea behind the approach is in principle supported by Munday, who argues that '[comparing] two translations of the same ST, even though they are into different TLs, allows some triangulation of findings and helps to avoid jumping to conclusions based on a single isolated study' (Munday 2012:185). In other words, data from one target language can in principle be put together with that from another (or others) to form more reliable conclusions about translator behaviour.

Because the language-specific data is easily accessible, it is a simple matter to monitor what is happening in the various translated versions if the need arises—for example, to ensure that no one target language is skewing the results unduly. In addition, adopting a multilingual approach does not mean ignoring all the intrinsically interesting language-specific detail, as this comes up constantly in the discussion of examples.

The approach takes some of its inspiration from the thinking behind the translation universals but is not directly related to them and does not aim to make a contribution to that line of research. On the other hand, my hope is that the list of procedures that I am proposing in Section 7.1.3 will possess a certain degree of cross-lingual applicability.

At the same time, since all the translations are produced approximately simultaneously and the possibility of influence from one international edition to another is probably remote, the 'multilingual paradigm' is likely to be significantly different from that of retranslation, not least since the status of a target text as a retranslation is determined by the 'prior existence of an initial translation of a given work into a given language' (Deane-Cox 2014:1). An alternative multilingual approach—and not the one taken in this study—would have been an explicitly comparative one in which the output of each language's group of translators would have been contrasted. This would have more in common with most research into retranslation, but with the identity of the target language being at issue in the former and matters such as the time that elapses between successive retranslations coming into play in the latter.

3.4 Conclusion

There are many different approaches to the theoretical study of metaphor, although the most popular one among researchers is currently the conceptual metaphor theory. This theory has developed considerably since it was first proposed in 1980, and it is in one of its later forms, which allows more of a place for interlingual and intercultural variation in metaphorical language, that I am using it in my research. Much has been written over the last few decades on metaphor in translation, and the idea of utilising concepts from metaphor theory to help reflect on translation procedures is one that I consider to be very valuable. The research that forms the basis of this book draws on considerable quantities of data and is being pursued along the lines of an approach that I have developed that places this data's multilingual nature at its centre.

Notes

1. It is generally held that source and target domains are distinguished from each other in terms of their degree of abstractness (the source being less abstract than the target), although an alternative view is that of Grady (1997), who argues that, in fact, the distinction is along the lines of relative subjectivity (determined, for example, in terms of physical perception) rather than their relative levels of abstraction (see Evans & Green 2006:304).
2. The abbreviations SL and TL stand for 'source language' and 'target language' respectively. Similarly, ST and TT denote 'source text' and 'target text'.
3. Parts of this section, and also of Sections 3.2.1.2, 3.2.1.3 and 3.2.2.5, as well as a brief passage in the introduction, have been adapted and reprinted in part with permission from 'Translation Studies and Metaphor Studies: Possible Paths of Interaction between Two Well-Established Disciplines', in Donna R. Miller & Enrico Monti (eds) *Tradurre Figure/Translating Figurative Language*, Quaderni del Centro di Studi Linguistico-Culturali, Atti di Convegni CeSLiC—3, Bologna: AMSActa, 53–65. Copyright 2014, Bononia University Press (Shuttleworth 2014).
4. Shuttleworth (2014), which was referenced in the footnote at the start of Section 3.2.1.1, also contains a section in which I discuss the importance of maintaining a two-way dialogue between the two disciplines.
5. What precisely is meant by boldness will be discussed briefly in Section 5.2.2.
6. The sections of this work that deal with metaphor in translation are a revised version of Snell-Hornby (1983).
7. This idea is very much expanded in the section on 'Polyseme' in Dagut (1978:103–9).
8. When she uses the term *non-literary*, Pisarska is actually referring to a corpus of seven books relating to the fields of history, anthropology, economics and psychology.
9. This figure has been reproduced with the kind permission of Mary Snell-Hornby.

4 Macro-Level Metaphors

We start our analysis of the six different parameters that are covered in the book by focusing on existence categories and mappings. They are being considered together because they can both be thought of as providing ways of categorising metaphor semantically on a high level. Of the two, the former is probably more in line with traditional ways of classifying metaphorical expressions, while the latter is a central concept in the cognitive theory of metaphor.

4.1 Existence Category

For reasons set out at the beginning of Section 4.1.3.1, the discussion of existence categories is shorter than most of the five following chapter sections that are devoted to other parameters. Like the others, it will start by providing a theoretical overview of this way of categorising metaphor. It will then consider how the concepts that have been discussed are reflected in the English data, after which the remaining part of the section will look at translation procedures, illustrating each point with at least one example from the data.

4.1.1 Background

In view of the very large number of individual mappings contained in the data (this will form the content of Section 4.2), I believe it also to be sensible to relate the semantic content of the metaphorical expressions included in the data to a more limited number of categories (such as *humanisation* or *reification*) that exist at a higher level of generalisation. As stated at the beginning of the chapter, the resulting parameter, which I am terming 'existence category', is not one that forms a part of the cognitive theory of metaphor, and, presumably largely as a consequence of this, it has not received anything like the same degree of theoretical attention as the concept of mapping, in either metaphor research or translation studies.

That said, a number of scholars have proposed schemes for classifying metaphors in terms of their high-level semantic features that are of varying degrees of formality and complexity. Four of these are summarised in Table 4.1. To varying extents, these taxonomies—with the possible exception of Metalude—are unconsciously predicated on the idea of moving either up or down the 'Great Chain of Being', a culturally conditioned ordering of creatures and other entities based on the idea that everything in the universe is arranged hierarchically, with humans near the top, followed by animals, then plants and finally inanimate objects and substances right at the bottom (see Lovejoy 1936, and also Lakoff & Turner 1989:166–9; Peña Cervel 2003:37–8; Norwick 2006:343).

Described by its authors as 'anything but a finished product', the Master Metaphor List is a partial compilation of material drawn from published sources, student papers and research seminars (Lakoff, Espenson & Schwartz 1991:1). Besides the fact that it contains a total of 69 subcategories, the general categories used by the Master Metaphor List are not suited to our present purposes: of the four categories, one would expect 'emotions' to be only marginally represented, given the subject matter of the corpus, while 'other' is too general to be of significant use in the classifying process.

Table 4.1 Different Schemes for High-Level Semantic Classification of Metaphor

The 'Master Metaphor List' (Lakoff, Espenson & Schwartz 1991)	*Goatly (1997:46–57)*	*Charteris-Black (2004:21)*	*Metalude (Goatly & LLE Project 2005)*
1. event structure (29 subcategories) 2. mental events (18 subcategories) 3. emotions (6 subcategories) 4. other (16 subcategories)	1. general reifying 2. specific reifying 3. animising 4. personifying 5. materialising abstract process 6. process = process 7. object = object	1. reification 2. personification 3. depersonification	Target categories: 1. Activity and Movement 2. Humans, Human Senses and Society 3. (Living) Things and Substances 4. Values, Qualities, Quantities 5. Emotions, Experience, Relationships 6. Thinking, Communication Source categories: A. Things and Substances B. Human/Animal Body and Senses C. Activity and Movement D. Space and Place

The principles on which Goatly's classification is based are much more in line with our purposes, although with 20 subcategories in all, it is perhaps also too intricate to be used to analyse a large quantity of data in an unmodified form. In addition, it is interesting to note that the author distinguishes both 'general' and 'specific' reification (1997:46–51), the former being reserved for lexical items that 'seem to apply very generally, that is to a large number of abstract entities' (1997:46), such as *emerge*, *exhibit* or *possess* (1997:47–50).

I have included Charteris-Black's three categories in order to illustrate the different types of semantic tension that are caused by a metaphor (2004:21). They are intended as a means of describing a particular metaphor's linguistic rather than pragmatic or cognitive characteristics (2004:21) and are not presented as an exhaustive taxonomy.

Metalude is based on the idea that every metaphorical expression can be assigned to an inter-domain mapping linked to a particular square on a grid representing the intersections between six target-domain and four source-domain groupings. Thus, for example, square 2D (target: Human Senses and Society; source: Space and Place) contains mappings such as SOUND FREQUENCY IS HEIGHT (e.g. 'a very deep voice') and HUMAN IS BUILDING (e.g. 'built-in intelligence').

Even though Metalude (Goatly & LLE Project 2005) represents a highly innovative refinement of the earlier list presented by Goatly, it is not being used here since the number of categories that it distinguishes (a total of 20, squares 1A, 2C, 3C and 3D being empty) would once again make its use too complex for my purposes.

The remainder of Section 4.1 will propose and apply a system that builds on the aforementioned but that closely reflects the subject matter of the project data.

4.1.2 *How These Concepts Feature in the Data*

As far as I am aware, no study has so far looked at metaphor in translation from this theoretical point of view. As with the other parameters, it is hoped that some valuable insights will be obtained from correlating the different categories with the particular approaches opted for by translators.

The classification scheme I have arrived at is based on that of Goatly, but it consists of only 9 categories rather than the original 20 (and no subcategories). Some of these originals have been removed, and a number of my own have been added; for ease of analysis, some of the more specific categories have been merged. Also, his distinction between general and specific reification is not being used. This is because in my view, it would need to be mirrored in at least one other category that I identify (concretisation), thus increasing the number of categories. The distinction between 'general' and 'specific' is in any case broadly reflected in the conventionality parameter, and the presence of more 'general' reification often results in a more generalised mapping being posited.

All the different existence categories that can be identified in the data are described below. Categorisation takes place on the mapping level where possible, but on that of individual metaphorical expressions where this proves necessary.

4.1.2.1 *Abstractive Transformation*

This concept accounts for mappings from one abstract concept to another (e.g. *feeding the desire to seek the drug* (Nestler & Malenka 2004:84)—the mapping here being DESIRE IS HUNGER). Abstractive transformation generally entails a degree of familiarisation.

4.1.2.2 *Concretisation*

This category encompasses mappings from one inanimate object to another inanimate object or, rarely, from an inanimate object to an abstract concept that make the topic of a metaphor easier to understand by bringing it closer to one's everyday experience (e.g. *transporter protein* (Nestler & Malenka 2004:82), *interchangeable DNA parts* (Gibbs 2004:75) or *Life, version 2.0* (Gibbs 2004:76), the mappings here being PROTEINS ARE TRANSPORT, GENETIC MATERIAL IS MACHINERY and GENETIC MATERIAL IS SOFTWARE, respectively). Sometimes—in the case of terminology, for instance—the concretisation is concealed by the fact that the vehicle is derived from another language (see Section 1.2.3).

4.1.2.3 *Dehumanisation*

This involves mappings from a human to a non-human entity (e.g. *Freud's concept of man-as-animal* (Solms 2004:87) or *a new conception of 'what makes us tick'* (Solms 2004:84), which instantiate the mappings PEOPLE ARE ANIMALS and PEOPLE ARE MACHINES).

4.1.2.4 *Humanisation*

Mappings from a non-human entity (i.e. animals, objects or abstracts) to a human one are placed in this category (e.g. *if glia could chatter* (Fields 2004:56), *chemical messenger molecules* (Fields 2004:55) or *a 'Goldilocks' genetic circuit* (Gibbs 2004:78), relating to the mappings NEURONS ARE COMMUNICATORS, NEUROTRANSMITTERS ARE MESSENGERS and CERTAIN GENETIC CIRCUITS ARE GOLDILOCKS). This phenomenon is sometimes referred to as anthropomorphism (see Kurth 1995:102 and Aleksander 2000:2).

4.1.2.5 *Humanotypic Transformation*

This term denotes mappings from one class of human to another (e.g. *researchers mapped the regions of the brain* (Nestler & Malenka 2004:81)—the mapping here being RESEARCHERS ARE CARTOGRAPHERS).

4.1.2.6 Personification

Mappings from abstract concepts or inanimate objects to sentient beings, whether or not these are specifically non-human (e.g. *neurons were free to communicate* (Fields 2004:55) or *the ongoing struggle between the ego and id for dominance* (Solms 2004:85), derived from the mappings NEURONS ARE COMMUNICATORS and THE EGO AND THE ID ARE BATTLING PARTIES) are distinguished from those that unequivocally involve humanisation.

4.1.2.7 Plantification

The concept of plantification encompasses mappings from non-plant to plant entities (e.g. *twiglike receptor, or dendrite* (Fields 2004:55), where *dendrite* derives from Greek *dendritēs* 'pertaining to a tree' and the mapping is NEURONS ARE TREES).

4.1.2.8 Processual Transformation

This term refers to mappings from one process or activity to another process or activity (e.g. *Ben Kater . . . defused that suspicion* (Fields 2004:57)—the mapping here being REFUTING IS DEFUSING). This category is sometimes associated with mappings on a higher level of generalisation, where mapping the process rather than the agents can enable one to avoid implausible humanisations (e.g. *des canaux ioniques qui permettent aux neurones de produire des potentiels d'action [some ion channels that permit neurons to fire action potentials]* (Fields 2004:59), where it is intuitively more reasonable to verbalise the mapping as the general TO MAKE SOMETHING POSSIBLE IS TO PERMIT IT rather than the highly specific and awkward-sounding ION CHANNELS ARE HUMANS).[1]

4.1.2.9 Reification

The category of reification is formed of mappings from abstract entities (or, occasionally, processes) to inanimate objects (e.g. *unpalatable information* (Solms 2004:85), *the evolutionary ladder* (Fields 2004:61) or *this process was a black box*, with the mappings here being UNPLEASANT IS UNPALATABLE, EVOLUTION IS A LADDER and UNCOMPREHENDED PROCESSES ARE HIDDEN FROM VIEW).

Interestingly, all nine of these categories reflect metaphor's explanatory function rather than its potential for mystification. A number of other theoretically possible categories turned out not to be represented in the data; these include depersonification, animisation, abstraction, kinetisation (i.e. mappings from a change, process or activity to a movement) and mappings from a process to an event, or from a state to a thing.

4.1.3 *Translation Procedures and Analysis of Examples*

The time has now come to take a look at what happens to metaphor in translation in the context of this parameter. Because of the relatively large number of parameter values, I will not be presenting a detailed category-by-category analysis of translation procedures but will content myself with a) exemplifying some of the trends that are demonstrated by Table 4.2 and by way of further illustration b) presenting a brief study of humanisation and personification, and specifically of expressions within these categories that involve cognitive functions of a relatively general nature.

4.1.3.1 Numerical Background

Table 4.2 sets out the relative frequency of the different existence categories in English and the six target languages. Like the similar tables that follow in Sections 5.1.3.1, 5.2.3.1, 6.1.3.1.1 and 6.1.3.2.1 (Tables 5.1, 5.2, 6.2 and 6.4), Table 4.2 shows how the source-language expressions belonging to a particular

Table 4.2 A Comparison of the Numbers of Metaphorical Expressions in Each Existence Category in the Source and Target Texts

Source-language existence categories (numbers of source-language expressions shown in brackets)	*Target-language existence categories (all six target languages taken together)*								
	Abstractive transformation	*Concretisation*	*Dehumanisation*	*Humanisation*	*Humanotypic transformation*	*Personification*	*Plantification*	*Processual transformation*	*Reification*
abstractive transformation (2)	1	0	0	0	0	0	0	0	1
concretisation (136)	0	493	0	1	0	5	0	0	0
dehumanisation (2)	0	1	7	0	0	0	0	0	0
humanisation (44)	0	5	0	138	0	9	0	0	1
humanotypic transformation (3)	0	1	0	0	5	0	0	1	0
personification (31)	0	2	0	6	0	101	0	0	0
plantification (13)	0	2	0	0	0	0	37	0	0
processual transformation (9)	0	1	0	0	0	0	0	14	1
reification (28)	0	1	0	1	0	0	0	3	70

category are rendered in translation; it enables us to identify any particular categories whose members are not simply routinely replaced by equivalents from the corresponding target-language category and thus provides us with fairly quick access to data from which we can infer the likely nature of any characteristic parameter-specific procedures that might exist.

The numbers of metaphorical expressions in each target language are relatively low since the table does not include those that are either removed, omitted or otherwise absent. Given the fact that the overwhelming majority of metaphorical expressions that survive the translation process retain their membership of a certain category in translation, we can infer that no particular category presents translators with special problems or difficulty. However, on the other hand, it is interesting to note that it is some of the less common categories that attract greater diversity in translation approach (although there may be no significance in this fact). We can also infer from the table where large-scale removals or omissions have taken place: for example, given that the English data contains two examples of abstractive transformation, we would expect to see up to 18 reported in the 9 target-language columns in that row, so the fact that only 2 are present indicates that 16 instances have not been retained as such in translation.

4.1.3.2 Examples of the Different Procedures

As stated at the beginning of Section 4.1.3, the analysis of procedures will not be as detailed as those of some of the other parameters.

i) Let us first consider the commonest category, that of concretisation. To start with, let us look at one long example and then three much shorter, language-specific ones. After that, a much shorter second section will be devoted to one particular procedure that seems to be followed in a small number of cases.

a) The widespread preservation of this large category in translation means that there are not a few instances in which a concretised metaphorical expression features in five or six target versions, as is the case in the following example:

English:	two artificial genetic **'letters'** in addition to the four that appear in life as we know it. (Gibbs 2004:76)
German:	wies neben den üblichen vier genetischen »**Buchstaben**« (Nucleinsäurebasen) . . . zusätzlich zwei künstliche Basen auf. (Gibbs 2004:69) [besides the usual four genetic **'letters'** (nucleic acid bases) . . . displayed in addition two artificial bases]
Russian:	ДНК, содержащую кроме четырех известных **букв** генетического алфавита еще две. (Gibbs 2004:48)

[DNA containing besides the four familiar **letters** of the genetic alphabet two further ones]

Polish: prócz czterech naturalnie występujących w przyrodzie „**liter**" dodatkowo dwie sztuczne. (Gibbs 2004:60)
[besides the four naturally occurring **'letters'** two further ones]

Italian: con due «**lettere**» genetiche artificiali in più oltre alle quattro che compaiono nella vita «naturale». (Gibbs 2004:79)
[with two **'letters'** that are genetically artificial in addition to the four that are present in 'natural' life.]

Chinese: 除了包含已知構成生命的A、T、G、C四種遺傳**字母**外，還具有另外兩種人造的遺傳**字母** (Gibbs 2004:88)
[as well as containing the four known constituent **letters** of heredity A, C, G and T, it also possesses two further artificial heredity **letters**]

As stated in the table, there are 493 instances in which a concretisation is retained as such. This particular example, which is taken from 'Synthetic Life', has been chosen as it is representative of a group of expressions that all relate to the GENETIC MATERIAL IS TEXT mapping. The metaphorical words that are involved—*letters* and, elsewhere, *alphabet*, *sentence*, *word* and *punctuation*—are exegetical in nature (see Section 5.1.2.3), although they have by now become more or less part of the informal scientific terminology.

What we see here is that the English metaphorical expression is retained in all five target languages in which a translation exists. Moreover, this general pattern of retention (albeit with occasional removals or omissions) is repeated in the other three examples that also involve *letters* and, of course, in a more generalised manner across many other translations of the 136 English concretisation expressions. That said, I would not expect what occurs in all these examples to be simply a matter of 'retain means retain': there is in fact some variation in what retention can actually entail, as will be seen in some detail in Sections 6.1.3.1.3 and 6.1.3.2.2.

Parenthetically, it is interesting to note that the only slight deviation from this general use of retention in these GENETIC MATERIAL IS TEXT expressions can be found in the Chinese data, where there is no mention of 'letters' in either of the other examples:

English: Re-creating a virus **letter-for-letter** (Gibbs 2004:80)
Chinese: 重新打造<u>一模一樣</u>的病毒 (Gibbs 2004:92)
[recreating <u>the identical</u> virus]

English: an alphabet of just four **'letters'** (Freeland & Hurst 2004:86)
Chinese: 由四種<u>核苷酸</u>所寫成 (Freeland & Hurst 2004:78)
[consisting of four <u>nucleotides</u>]

In addition, in the following example, *letters* is retained while *alphabet* is removed:

English: **a six-letter** genetic **alphabet** (Gibbs 2004:80)
Chinese: 一套由**六個字母**構成的基因**語言** (Gibbs 2004:93)
[a genetic **language** consisting of **six letters**]

What might be the cause of this difference in approach? As I understand it, the concepts of the genetic *letter* and *alphabet* are just as much part of Chinese scientific terminology as they are of that of many other languages, and yet, on the other hand, most educated native speakers of Chinese are likely to feel more in tune with the broadly logosyllabic Chinese writing system (while also being familiar with the Roman alphabet, of course), so maybe it is this that is having an unconscious effect on the translator and producing an effect that is different from that seen in the five alphabet-based languages. If true, this is a fascinating example of how a deep-seated cultural factor can influence a series of translation decisions, although, of course, it is impossible to be sure that this is anything more than an anomaly without examining more data.

b) In a very subsidiary manner, it is also worth looking at an example in which a concretisation is shifted to a personification:

English: When an axon **fires** (Fields 2004:56)
Chinese: 當軸突**激發**時 (Fields 2004:37)
[When an axon **is excited**]

Three of the other instances also involve the *firing* of neurons, and in each case, the Chinese translation includes the word 激發 ('excite'), while the last one entails a similar kind of rewording in French. Personally, however, I do not find these particularly interesting from a translation point of view, as I believe the translator's decision may simply represent an attempt to conform to the collocational norms of the target language.

ii) Table 4.2 indicates that nine humanisations are translated by a personification and six personifications by a humanisation.

a) The following will serve as an example of the former:

English: if glia could **chatter** (Fields 2004:56)
Polish: **porozumiewanie się** komórek neurogleju (Fields 2004:28)
[glial cells **making themselves mutually understood**]
French: si les cellules gliales **communiquent** (Fields 2004:57)
[if the glial cells **communicate**]
Italian: se fossero state in grado di **comunicare** (Fields 2004:50)
[if they had been capable of **communicating**]

This categorisation is based on the assumption that, while *chattering* as an activity characteristic to humans, *communication* is a process that is indulged in much more widely. Possible motivations for this procedure are not very obvious.

b) Shifts in the opposite direction (i.e. personification to humanisation) are exemplified by the following:

English: how exactly would glia be affected by what they **heard**? (Fields 2004:56)

Russian: как влияют на работу глиальных клеток «**подслушанные**» ими нервные сообщения? (Fields 2004:25)
[how is the work of the glial cells affected by the neural communications that they **'overhear'**?]

In this case, unlike the original, the translation involves an action that is specific to humans. Of the five examples, this one is perhaps noteworthy since it seems to result from a specific decision on the part of the translator. In some of the ones that have not been presented, on the other hand, the shifts that we see (e.g. *perceive* becoming *consider*) may simply be the side effects of almost random rewordings—something that also applies to the humanisation to personification shifts discussed in point a) earlier.

What is not certain—particularly given the very small number of examples available—is whether it would be possible to talk about a 'preference' for humanisation or personalisation either in the practice of translators working into a particular language or in terms of the metaphorisations existing in the language as a whole. My suspicion is that while the former is probably unlikely, it is possible that the latter tendency could indeed be observed at least partially and maybe on a domain-by-domain basis.

iii) Humanisation is shifted to concretisation in five examples including the following:

English: the influx in glial cells would also activate enzymes that would **marshal a response** (Fields 2004:57)

Italian: il flusso di ioni nelle cellule gliali attivasse enzimi destinati a **stimolare una reazione.** (Fields 2004:51)
[the flow of ions in the glial cells activated enzymes destined to **stimulate a reaction**]

Once again, the nature of some of the examples suggests these rewordings were caused by factors other than those relating to this parameter.

iv) There are three examples of reification becoming processual transformation:

English: cognitive control mechanisms that **underpin** normal reality monitoring (Solms 2004:87)

German: kognitive Steuerungsmechanismen . . ., die normalerweise eine realitätsgerechte Einstellung **gewährleisten**. (Solms 2004:81)
[cognitive control mechanisms . . . that normally **ensure** an attitude that is appropriate to reality]

Russian: механизмы когнитивного контроля, **обеспечивающие** нормальное восприятие реальности (Solms 2004:59)
[cognitive control mechanisms that **ensure** a normal perception of reality]

French: les mécanismes cognitifs de contrôle **assurant** normalement le suivi de la réalité (Solms 2004:80)
[cognitive control mechanisms that normally **ensure** reality monitoring]

These are the only examples of this procedure, and they seem to be motivated by the fact the specific English mapping A SUPPORT IS A PIN gives way in these three target languages to the more general ENABLEMENT IS ASSURANCE.

Table 4.2 reveals a number of more minor procedures that are exemplified by as few as a single example, and these are not reported here. Likewise, some of the more frequent procedures are also not discussed for the reason given at the beginning of Section 4.1.3. However, the impression that may have been gained from some of the examples presented is that this parameter do not seem to be the focus for translation decisions as the (small number of) shifts in this area seem to a large extent to be a knock-on effect from changes the motivation for which might be totally unrelated.

4.1.3.3 Humanisation and Personification in Translation

One of the features of the popular scientific texts that we can observe at this level is the relatively large number of humanising and personifying expressions, which include, more specifically, verbs that denote different conscious cognitive functions (e.g. *know*, *sense*, *monitor* and *detect*). The various metaphorical expressions of which they form a part are associated with a range of different mappings. At the level of the existence categories, however, some of these are classified as personification rather than humanisation because they do not represent an exclusively human mental faculty. The question can therefore be asked whether any consistent translation shifts might be observed at my data's highest level of generalisation.

The overall number of instances of humanisation in the English data is reduced by nearly half in the target languages through removal and omission, as is that of personification. Where there is a shift to another existence category, as reported in the previous section, personification normally becomes humanisation and vice versa:

English: neurons and glia **engage in a two-way dialogue** (Fields 2004:55)
Polish: neurony i komórki glejowe **komunikują się ze sobą** (Fields 2004:27)
[neurons and glia cells **communicate with each other**]

As in an earlier example, here the specifically human activity of conducting a dialogue is mutated into one that can be performed by any sentient being.

There are two tentative generalisations that it is perhaps worth making concerning particular shifts that can be observed in the translation of the handful of examples of personification that involve conscious cognitive functions. First, there is a tendency to replace stylistically relatively neutral verbs with more specific ones:

English: glia **sense** axon activity (Fields 2004:57)
German: dass Gliazellen neuronale Impulse **wahrnehmen** (Fields 2004:50)
[that glial cells **are aware of** neuronal impulses]
French: les cellules de Schwann **détectent** l'activité des neurones (Fields 2004:58)
[the Schwann cells **detect** the neuron activity]
Italian: la glia **percepiva** l'attività degli assoni (Fields 2004:51)
[the glia **perceive** the axon activity]
Chinese: 神經膠細胞 . . . **感知**軸突的活動。(Fields 2004:38)
[glia . . . **perceive** axon activity.]

All the examples of the verb *sense* in the various target languages that are not either removed or omitted conform to a similar pattern: in some 16 cases, the expression selected in the target text is more specific, and sometimes also stylistically more sophisticated, than that found in the original.

The second generalisation that can be made is that there is a certain preference for passive-equivalent constructions (as in the use of a reflexive verb with a passive meaning) in Russian and Polish:

English: they [i.e. computer predictions] **ignore** the wobble phenomenon (Freeland & Hurst 2004:90)
Russian: в них не учитывается феномен «качания» (Freeland & Hurst 2004:60)
[in them no account is taken of the 'wobble' phenomenon]
Polish: **nie bierze się w nich pod uwagę** zjawiska wahadła (Freeland & Hurst 2004:62)
[in them no attention is paid to the pendulum phenomenon]

In all, there are seven examples of this. At this stage, because there are comparatively few significant clusters of relevant metaphorical expressions in the data, I have only been able to conduct a relatively limited investigation, and so it would be premature to draw any major conclusions. However, the examples that I have analysed provide a tantalising glimpse of the interesting possibility that at least some apparent tendencies in translation can be observed at this level of generalisation. Clearly, this is an area that merits further investigation.

4.1.3.4 Conclusion

This study has revealed certain interesting phenomena that may merit further investigation, such as the handling of verbs denoting a range of conscious cognitive functions within the context of humanising or personifying metaphors, but at the same time, there is the sense that some of the shifts that have been observed may be purely fortuitous when seen from the point of view of this parameter. Existence categories could in fact be a fruitful level on which to pursue this kind of study, partly because there are some signs that it may be possible to detect certain tendencies in the way in which translators tackle metaphorical expressions belonging to particular classes. There is great scope for further research—for example, within thematic areas such as genetic material and the modelling of machines as people—once more data becomes available.

4.2 Metaphor as Mapping[2]

Because of the large number of mappings identified in the data, this section will also not attempt anything like an exhaustive study of the possible translation routes that are opted for by translators, as seen from the point of view of this parameter. What it does contain, on the other hand, is a discussion of the mappings that occur most frequently in the English data and how instantiations of them are translated. This is followed by a closer look at two of them.

4.2.1 Background

The mapping arguably represents the most fundamental way of categorising metaphors and metaphorical expressions within the conceptual metaphor theory. In addition, in common with the existence categories, which we have already looked at, and unlike most of the other parameters, it relates directly to a metaphor's subject matter rather than some more theoretical characteristic.

I prefer the word 'mapping' to the more usual term 'conceptual metaphor', as mappings quite clearly underlie all metaphorical expressions—not only instantiations of conceptual metaphors but also image metaphors.

Typically, researchers will propose a particular formulation (or 'verbalisation') for a mapping on the basis of their understanding of the data. It goes without saying that finding the right verbalisation for a mapping is not always easy and generally involves an element of subjectivity. The way it is verbalised needs to encapsulate its metaphorical force and to be located at what the researcher judges to be the most appropriate level of generality on the basis of its apparent range of applicability. Humanisations, for example, can range in generality from the very broad BRAIN CELLS ARE SENTIENT to the highly specific NATURE IS A BLIND DESIGNER, depending on the mapping's precise contextual motivation. More specific mappings (e.g. NATURE IS A BLIND DESIGNER and NATURE IS A SOFTWARE DESIGNER) can coexist with more general ones (e.g. NATURE IS A DESIGNER and NATURE IS A HUMAN), inheriting some of their entailments from them. In addition, where related mappings for a single target domain appear to exist, individual metaphorical expressions need to be assigned carefully to the most appropriate one. For example, the data allows us to distinguish between GENETIC MATERIAL IS TEXT and GENETIC MATERIAL IS LANGUAGE, and so decisions need to be taken for expressions containing lexical items such as *editing out*, *libraries* and *transcript* (all GENETIC MATERIAL IS TEXT) and *acid meanings*, *nonsense* and *synonyms* (all GENETIC MATERIAL IS LANGUAGE).

By convention, the citation form of a mapping is usually 'A IS B'. Although some writers are happy to use other, non-copula structures (see, for example, Gibbs 1999a:31; Al-Harrasi 2001; Farrell 2004:48), I am adhering to this form in all the mapping verbalisations that I discuss.

Some mappings do not really exist beyond the description of a single lexical item or very limited group of items (for example, A THEORY IS A FRAMEWORK and a 'new intellectual *framework* for psychiatry': Solms 2004:88). It also sometimes happens that a single expression (or, more typically, a passage containing multiple metaphorical expressions) can be attributed to more than one mapping because of the sheer difficulty of determining the main entailment that is intended (if there is just one) and therefore verbalising the mapping in an appropriate manner. For example, 'frequently jumbling words together into a meaningless word *salad*' (Javitt & Coyle 2004:50) appears to be an instantiation of a mapping such as WORDS CAN BE MIXED LIKE SALAD or INCOHERENT SPEECH IS MIXED LIKE A SALAD, but probably not of LANGUAGE IS A PREPARED DISH. Indeed, there are sometimes several parallel ways of seeing the same concept, each reflected in a separate mapping (for instance, GENETIC MATERIAL IS TEXT, GENETIC MATERIAL IS CODE, GENETIC MATERIAL IS DATA, etc.).

4.2.2 *How These Concepts Feature in the Data*

This parameter is very different from the others in that it has more than 100 possible values, whereas each of the others possesses fewer than 10. Throughout my research, I use it tacitly as the main categorising feature.

This is not only because a change of mapping—and consequently of the ideational content of an expression—is more likely to result from a conscious decision on the part of the translator than a modification in any of the other parameters but also because the possibility of discovering evidence that translators demonstrate an awareness of these large metaphorical structures and are sensitive to cross-lingual and cross-cultural differences in how the mappings that describe a particular area are delineated is one that researchers in this area seem to find somewhat alluring.

The tentative nature of some of the more specific mappings should be very clear, not least given the fact that many of them are only represented by a single example. This creates the frequent need to verbalise a mapping for single expressions, which causes new mappings to be proposed in a fairly open-ended manner. This open-ended coining of mappings is not a problem as long as they are only viewed as convenient labels and their precise configuration is not expected to bear any significant conceptual or theoretical load, and providing that the process is not allowed to become an end in itself. My original set of mappings, it should be noted, are attempts to characterise a single expression or group of expressions identified in the English texts, and a modified expression in one or the other target text will sometimes require a new mapping to be posited to account for it. In addition, as pointed out by Dobrovol'skij and Piirainen, there is the possibility of postulating 'various alternative conceptual metaphors for the same metaphorical expression' (2005:129).

Although it has been claimed by Lakoff that mappings tend to be at the superordinate level (1993), Kövecses maintains that they can occur at different levels (2006:129–30). This latter insight is borne out in this study, where I have found the most expedient approach to be to posit mappings at the level of generality required by the expression, if it is a stand-alone one, or to reflect the intuitive 'groupings' that suggest themselves and then verbalise accordingly. In other words, I have tried to place mappings at a sufficiently high level of generalisation to account for all expressions that belong to an obvious grouping of expressions and at a sufficiently low level to permit differentiation between clearly different but related sub-groupings. It is for this reason that the various humanisations of nature are verbalised at different levels of specificity, giving rise to a kind of hierarchy of mappings: NATURE IS A HUMAN—NATURE IS A DESIGNER—NATURE IS A BLIND DESIGNER and NATURE IS A SOFTWARE DESIGNER. The humanisations towards the specific end of the spectrum are frequently specialised—in the sense that they assign to nature a specific human role or even profession. Mappings at the general end, as well as other generalised ones such as GENES ARE HUMANS and BRAIN CELLS ARE SENTIENT, on the other hand, are allowed to remain at that level of generality because, in some cases, there is little to be gained by increasing the level of specialisation and, in others, there are too few examples available in the corpus to make this necessary. In the absence of any need for greater specificity, such mappings effectively draw on what is a universal physical, cognitive or sensory human capability, such as the ability to hear or to find one's way. Some mappings are expressed in very specific terms—for example, SATISFYING AN ADDICTION IS HITTING A TARGET or UNCOMPREHENDED PROCESSES ARE HIDDEN FROM VIEW—although this is often caused by the need to adhere to the A IS B format. (An interesting question to consider, within the context of metaphor rather than translation research, would be whether this insistence on a rigid syntax for expressing mappings placed an unnecessary constraint on our understanding of the relationship between source and target domains in a mapping.)

The *word salad* example discussed in Section 4.2.1 was a case that presented genuine difficulty when I was considering the best verbalisation for the mapping that it instantiates. However, it was by no means the only such problem that had to be faced, with at least a dozen other such instances having to be confronted. Mappings should by no means be thought of as in any way immutable or definitive, although one that has been carefully chosen will have a high measure of explanatory power and will permit us to classify metaphorical expressions in a manner that is semantically and cognitively convincing.

4.2.3 *Translation Procedures and Analysis of Examples*

If individual metaphorical expressions represent the micro-level, mappings are structures that belong to the macro-level. As discussed in Section 4.2.2, it is often necessary to posit a new mapping to describe a single metaphorical expression. However, mappings that reflect a larger number of metaphorical expressions will of course have greater explanatory power, while the most widespread mapping within the corpus accounts for as many as 21. One of the interesting things about the data in the corpus is the different degrees to which the various mappings are reflected in English and each of the six target languages. The aim of this section is to investigate what if any translation trends can be observed on the mapping level, focusing to a large extent on the most frequently occurring ones. Some mappings—such as GENETIC MATERIAL IS MACHINERY and GENETIC MATERIAL IS CODE—are very stable in translation. But what occurs when problems do arise, and what is the nature of such problems? More generally, is it possible to any extent to generalise about what happens to mappings in translation?

4.2.3.1 *The Most Frequent Mappings*

All in all, I have categorised the English metaphorical expressions into 103 different mappings; of these, the most commonly occurring ten are as in Table 4.3.

Furthermore, within the 103 English-language mappings at least three clusters of thematically related ones are quite clearly delineated, centring

Table 4.3 The Most Frequently Occurring Mappings in the English Data

Mapping	*Number of expressions*
BRAIN CELLS ARE SENTIENT	21
GENETIC MATERIAL IS TEXT	16
NEURONS ARE TREES	13
THE NERVOUS SYSTEM IS CIRCUITRY	12
GENETIC MATERIAL IS MACHINERY	12
NEURONS ARE COMMUNICATORS	11
GENETIC MATERIAL IS CODE	11
GENETIC MATERIAL IS LANGUAGE	10
GENETIC MATERIAL IS SOFTWARE	9
NATURE IS A CREATOR OF CODE	9

around genetic material (eight mappings), nature (six) and neurons (six). The first two of these form the subject of the two Interludes later in the book, while the third is discussed in Section 4.2.3.2.2.

As it turns out, far fewer than the total of 103 different mappings contained in the English data can be found in each of the target languages (62 for German, 70 for Russian, 81 for Polish, 67 for French, 57 for Italian and 75 for Chinese), even counting the considerable numbers of new mappings that can be found in the target texts (the Polish translations contain around 20, for example). This striking circumstance can largely be explained by the fact that there are 64 English mappings that are only represented by a single example, and so can easily vanish in translation, depending on a single decision by the translator. Some examples of mappings that only occur in one or more target text will be provided in Interlude One.

What we observe overall is relative instability on both mapping and individual metaphorical expression levels. However, what is of greatest interest to us is whether the data indicates an overall tendency on the part of translators not to intervene on the macro-level or take global decisions regarding the acceptability (or otherwise) of a particular mapping in the target context. What evidence is there concerning this, one way or the other? Is it possible to observe an entire mapping being clearly shifted in one direction or another? Or is there rather a tendency for the individual expressions to become scattered when not retained? In order to investigate this we will now need to look at some individual mappings.

Among the 10 most frequently encountered mappings, there are no examples of one not being translated at all, and very few of one being seriously reduced in translation. These 10 mappings were listed in Table 4.3, but are now presented again in Table 4.4 for each language and overall:

Table 4.4 The Most Frequently Occurring Mappings in Each Language and Overall

Source-language mappings (numbers of source-language expressions shown in brackets)	*German*	*Russian*	*Polish*	*French*	*Italian*	*Chinese*	*Overall*
BRAIN CELLS ARE SENTIENT (21)	9	13	17	10	16	14	86
GENETIC MATERIAL IS TEXT (16)	9	14	15	8	15	8	77
NEURONS ARE TREES (13)	6	8	9	2	6	8	44
THE NERVOUS SYSTEM IS CIRCUITRY (12)	4	5	10	11	13	8	55
GENETIC MATERIAL IS MACHINERY (12)	8	11	11	1	10	8	53
NEURONS ARE COMMUNICATORS (11)	4	5	4	7	9	6	40
GENETIC MATERIAL IS CODE (11)	6	10	8	7	8	6	50
GENETIC MATERIAL IS LANGUAGE (10)	7	5	8	9	9	10	48
GENETIC MATERIAL IS SOFTWARE (9)	6	4	5	1	6	7	31
NATURE IS A CREATOR OF CODE (9)	3	3	1	6	3	6	25

This table indicates the relative frequencies of the ten most frequent mappings in each of the target languages. Not surprisingly, the order of frequency in the target languages, and overall, differs somewhat from that in the English texts and, as usually is the case, the number of examples of each mapping in the translations is generally lower than in the source texts. Two of the mappings will now be analysed in detail.

4.2.3.2.1 BRAIN CELLS ARE SENTIENT

The mapping at the top of this list, BRAIN CELLS ARE SENTIENT, encompasses a wide range of different kinds of personification, generally conveyed by a single word used in a figurative sense, as can be seen from the examples that follow:

> other glia **residing** at the axon's synapse (Fields 2004:59)
> more glia had lit up, because they had **sensed** the signals (Fields 2004:59)
> chemical signals that neurons did not **recognize** (Fields 2004:56)
> how exactly would glia be affected by what they **heard** (Fields 2004:56)
> Schwann cells **perform** myelination **duties** (Fields 2004:57)
> how an immature Schwann cell . . . **knows** which axons will need myelin (Fields 2004:59)
> glial cells merely **nurtured** them (Fields 2004:56)
> the overall **behavior** of nerve cells (Nestler & Malenka 2004:82–3)

Of the 21 examples, 16 are classified as discoursal and 5 as evaluative, which means that this mapping is used exclusively for non-core scientific purposes. Of the 21 examples, 20 are taken from the article 'The Other Half of the Brain' and 1 from 'The Addicted Brain'. To all intents and purposes, what happens to the mapping in each language is therefore in the hands of a single translator.

The following two examples focus on the idea of *sensing*, which is the item through which this mapping is chiefly instantiated in a total of five metaphorical expressions. As can be seen from these examples, translators are not always consistent in the way in which they translate individual metaphorical items. Indeed, it is sometimes difficult to identify an underlying pattern as to whether the appearance or non-appearance of a particular metaphorical expression in a particular target language can be attributed to purely metaphorical considerations. It is probably not always the case that translators are unhappy to replicate a particular metaphorical usage in a target text but, on the other hand, this circumstance does provide evidence that translators feel the need to aim for consistency in how they translate what they may well view as isolated expressions that are not central to the communicative force of the article. Thus, as we can see, in the following example,

English: more glia had lit up, because they had **sensed** the signals (Fields 2004:59)

Russian: свечение охватило еще большее количество глиальных клеток, **почувствовавших** сигналы нейронов. (Fields 2004:27)
[fluorescence took hold in an even larger number of glial cells that had **felt** the neurons' signals.]

French: de nouvelles cellules gliales s'illuminent, parce qu'elles ont **détecté** les signaux. (Fields 2004:60)
[some new glial cells light up because they have **detected** the signals.]

Italian: le cellule gliali fluorescenti erano aumentate di numero, indicando che i segnali le avevano raggiunte. (Fields 2004:53)
[the fluorescent glial cells were increased in number, indicating that the signals had reached them.]

the metaphoricity is preserved to varying degrees in the Russian and French, removed in the Italian and omitted in the German and Polish. (The Chinese was not available.) On the other hand, in this second example,

English: indicating they **sensed** the message (Fields 2004:60)

German: Sie hatten also die Nervenimpulse registriert (Fields 2004:52)
[So they had registered the nerve impulses]

Russian: признак того, что глиальные клетки отреагировали на это событие (Fields 2004:28)
[the sign that the glial cells had reacted to this event]

Polish: Oznacza to, że **wyczuwają** one wiadomość (Fields 2004:32)
[This indicates that they **sensed** the message]

Italian: Questa reazione indicava che **percepivano** il messaggio (Fields 2004:54)
[This reaction indicates that they **perceived** the message]

the metaphorical expression is preserved in Polish and Italian, removed in German and Russian and omitted in French. (Once again, the Chinese was unavailable.) In addition, throughout these and similar examples, we also observe some shifts in the level of formality, with the French translation of the first expression and arguably all those of the second placed on a higher stylistic level than that of the informal source expression.

Speaking from a multilingual perspective, it appears that this mapping survives well in the various target languages. This claim is made on the basis that at least six different translators working with different language pairs have all decided in most cases to translate relevant metaphorical expressions with target-language expressions that reflect the same mapping.

4.2.3.2.2 NEURONS ARE TREES

The mapping NEURONS ARE TREES is reflected in six English metaphorical expressions from 'The Other Half of the Brain' and seven from 'The Addicted Brain'. According to this mapping, the attributes and component parts of neurons are equated with those of trees on the basis of their appearance:

Each neuron has **a long, outstretched branch** (Fields 2004:55)
unneeded synapses **get pruned** (Fields 2004:61)
twiglike receptor, or dendrite (Fields 2004:55)
But those who become addicted to cocaine **sprout additional spines on the branches**, which consequently **look bushier** (Nestler & Malenka 2004:83)
twiglike receptor, or **dendrite** (Fields 2004:55)
signal-receiving **projections called spines** (Nestler & Malenka 2004:83)

This mapping is of course an image metaphor. This is in contrast to the previous mapping discussed and is unusual for the data, which contains only 21 image metaphor expressions, including the 13 reflected in this mapping. With 9 examples classified as exegetical (giving readers a clear idea of the appearance and structure of a neuron) and three as terminological, the overall function of this mapping is of surprisingly central importance and certainly not simply one of embellishment as one might have expected. (The remaining two examples are evaluative, however). It is at the same time worth noting that all of the 13 expressions contain a relatively rich image (e.g. numbers 1–4 of those listed above) except for two that involve terminology or other formal language (i.e. numbers 5–6 above). This proportion of rich images is considerably above the average for the corpus as a whole (see Section 6.1.3.2), although it is of course consistent with the nature of this mapping as an image metaphor.

In contrast to Section 4.2.3.2.1, and given that I will be returning to the NEURONS ARE TREES mapping in Section 6.2.3, I will here be focusing on specific target languages to provide a slightly more detailed look at a few examples similar in nature to those that I have just presented and representative of how translators working into most or all target languages handle expressions associated with this mapping. Here the Russian data represents a mixed bag of different procedures. Many of the renderings, however, show the translator choosing a more general word or phrase or a solution more loosely associated with the mapping NEURONS ARE TREES than what can be found in the source text:

English: to **buds** at its end (Fields 2004:55)
Russian: к расширенным участкам на его конце—аксонным терминалям (Fields 2004:24)
[to enlarged areas at its end—the axon terminals]

In this instance, the source-text metaphorical expression is removed, being replaced by a more general word (*buds* becoming *enlarged areas*). Like many of the modifications we see in the corpus, this appears to be an unforced change, as the translator could theoretically have used the Russian word for 'buds', *pochki*, but seemingly chose not to. The metaphorical term *aksonnye terminali* is then appended, adding a degree of explicitation and, arguably, raising the technical level of the sentence.

English: the Schwann cells **sprouted branches** (Fields 2004:61)
Russian: Тогда они дружно **выпустили в его сторону отростки** (Fields 2004:29)
[Then together they **put out shoots towards it**]

Here the unambiguous *sprouted* becomes the more general *put out*, and *branches* is changed to *shoots*. Quite clearly, we are still in the domain of plants, but not specifically that of trees.

English: But those who become addicted to cocaine **sprout additional spines on the branches**, which consequently **look bushier**. (Nestler & Malenka 2004:83)
Russian: У животных, пристрастившихся к кокаину, **шипиков на дендритах** гораздо больше (Nestler & Malenka 2004:55)
[Animals addicted to cocaine have far more **spines on their dendrites**]

Once again, *sprout* disappears (as does *bushier*), while *branches* this time becomes the less semantically immediate but more terminologically precise *dendrites* (for an explanation of this term, see Section 4.1.2.7).

English: **dendritic branches** with normal numbers of signal-receiving projections called spines (Nestler & Malenka 2004:83)
Russian: **дендриты** с обычным количеством шипиков—выростов, улавливающих нервные сигналы (Nestler & Malenka 2004:55) [**dendrites** with the usual number of spines—projections that catch nerve signals]

Here again the more semantically immediate word *branches* is deleted, leaving just the historically embedded metaphor *dendrity*.

So in the Russian examples, we see a definite weakening of the mapping, not least because of a tendency towards greater generalisation. This appears to be broadly typical of what we observe in the other target languages as well.

Finally, the mapping is best preserved in Polish, with only two examples removed and two omitted. What we see in this translation is a strong tendency to retain expressions essentially unaltered, although we can also find some evidence of a preference for similarly generalising solutions:

English: signal-receiving **branches** (Nestler & Malenka 2004:83)
Polish: receptorowe wypustki (Nestler & Malenka 2004:59) [receptor extensions]
English: to **sprout** additional **buds** (Nestler & Malenka 2004:83)
Polish: do tworzenia dodatkowych **węzłów** (Nestler & Malenka 2004:59) [to the creation of additional **nodes**]

In the two examples, the generalisation entails the removal and the reduction of the metaphorical expression respectively. Across all the target languages, these tendencies are widespread, indicating that, for whatever reason, the translators were maybe unsure about retaining this stark image in their target texts.

A clear conclusion that can be drawn is that most mappings seem to undergo some degree of reduction. At the same time, there is no clear shift from one mapping to another, but rather a kind of piecemeal dissipation. The examples we have looked at have also provided interesting data on how translators have responded to a number of different instantiations of an image metaphor, something to which we will be returning in Section 6.2.3.

4.2.3.3 Mapping Clusters

As stated earlier, an inspection of a complete list of mappings reveals that there are three significant thematically linked groups clustered around the concepts of genetic material, nature and neurons, consisting of eight, six and six (excluding NEURONS ARE TREES) English mappings, respectively, with a significant number of new ones added in the various target languages. In all, these account for a total of 65, 18 and 37 English expressions, respectively. For reasons of space, we will only be considering the first two of these three areas.

The first two of these three areas form the topics of the two interludes. As an example, however, I am presenting here in Table 4.5 a summary of the destination mappings of the source-language expressions associated with each of the eight mappings within the genetic material cluster:

It goes without saying that just because the mappings form a cluster, this does not mean that there is free traffic among all of them. On the other hand, there is a certain amount of shifting of expressions between the three thematically related mappings GENETIC MATERIAL IS CODE, GENETIC MATERIAL

Table 4.5 A Comparison of the Numbers of Metaphorical Expressions within the 'Genetic Material' Mapping Cluster in the Source and Target Texts

Source-language mappings within the genetic material cluster (numbers of source-language expressions shown in brackets)	*Target-language mappings within the genetic material cluster (all six target languages taken together)*							
	GENETIC MATERIAL IS TEXT	GENETIC MATERIAL IS MACHINERY	GENETIC MATERIAL IS CODE	GENETIC MATERIAL IS LANGUAGE	GENETIC MATERIAL IS SOFTWARE	GENETIC MATERIAL IS CIRCUITRY	GENETIC MATERIAL IS INSTRUCTIONS	GENETIC MATERIAL IS DATA
GENETIC MATERIAL IS TEXT (16)	60	0	1	2	0	0	0	0
GENETIC MATERIAL IS MACHINERY (12)	0	48	0	0	0	0	0	0
GENETIC MATERIAL IS CODE (11)	5	0	42	0	1	0	0	0
GENETIC MATERIAL IS LANGUAGE (10)	4	0	1	43	0	0	0	0
GENETIC MATERIAL IS SOFTWARE (9)	0	1	0	0	28	0	0	0
GENETIC MATERIAL IS CIRCUITRY (3)	0	0	0	0	0	10	0	0
GENETIC MATERIAL IS INSTRUCTIONS (3)	0	0	0	3	0	0	10	0
GENETIC MATERIAL IS DATA (1)	0	0	0	0	0	0	0	6

IS LANGUAGE and GENETIC MATERIAL IS TEXT, as well as between GENETIC MATERIAL IS INSTRUCTIONS and GENETIC MATERIAL IS LANGUAGE.

4.2.3.4 Mappings That Are Only Marginally Translated

There are a small number of mappings that do not survive the translation process in any language. Most of these (a total of 16) are only instantiated by a single expression, although two—DESIRE IS HUNGER and SATISFYING AN ADDICTION IS HITTING A TARGET—occur twice and there is one mapping—NEURONS ARE ENGINES—all seven expressions of which are almost invariably omitted or removed in each target language. The existence of such mappings, as well as the significant reduction that the mapping NATURE IS A CREATOR OF CODE that is discussed in Interlude One (as well as a small number of others) undergoes gives rise to the tantalising possibility that some mappings may be so closely linked with the resources of the source language that this reduction is all but inevitable.

The English mapping NEURONS ARE ENGINES appears to be inextricably associated with the verb *to fire*—e.g. 'when an axon **fires**' (Fields 2004:56), 'the **firing** of neurons' (Fields 2004:58). With few exceptions, each of the seven English expressions is either removed or omitted in German, Russian, Polish and French and modified to a different mapping in Italian and Chinese, as we see in the following example:

English:	necessary for **firing** impulses (Fields 2004:57)
German:	die Voraussetzung für Nervenimpulse. (Fields 2004:50) [the prerequisite for nerve impulses.]
Russian:	необходимый для генерирования нервных импульсов (Fields 2004:27) [essential for generating nerve impulses]
Polish:	niezbędnych do powstania impulsów (Fields 2004:30) [necessary in order for the impulses to arise]
French:	nécessaires à l'émission des potentiels d'action. (Fields 2004:58) [necessary for the emission of action potentials]
Italian:	per consentire la **scarica** degli impulsi. (Fields 2004:52) [to permit the **discharge** of the impulses.]
Chinese:	適合電脈衝的**激發** (Fields 2004:38) [suitable for the **excitation** of the electrical impulses]

In each of the first four of the translations, the translator chooses to remove or omit the metaphor entirely, while the NEURONS ARE ENGINES mapping is replaced by THE NERVOUS SYSTEM IS CIRCUITRY in the Italian and by NEURONS ARE SENTIENT in the Chinese. In point of fact, it is within the context of these two target-language mappings that most examples of retention occur. The fact is that because of the close association of the English mapping with the lexical item *firing*, if this item is replaced by its standard equivalent in one or other target language, an alternative mapping will generally be activated, as also occurs in isolated examples in Polish and French. This translation effect is, however, probably unremarkable.

In the case of all other mappings, at least one instantiation is translated into at least one target language. It is indeed worthy of note (even if it is unsurprising) that there are so few mappings of a relatively significant nature that are barely translated. Once again, this would appear to provide evidence for the hypothesis that when it comes to rendering metaphor, scientific translators do not adopt the kind of macro-textual strategy that would entail taking the naturalness of the mapping in the target language into account and do not generally introduce carefully considered modifications at the mapping level. (This may of course not be so in the case of literary translation, on the other hand.) However, an alternative hypothesis would perhaps be that factors such as suitability or appropriateness are relative in nature, and there is no such thing as a mapping that is completely alien to a particular language, even if it would not be the first choice of a native speaker. This is very likely to be the case, although is not mutually exclusive with the earlier suggestion. No generalisations are therefore possible regarding the nature of 'untranslatable' mappings, as there is no evidence that such a thing exists, except on the trivial level of mappings that are so specialised that they are only summoned up by the use of a single lexical item in a particular context in a particular language system.

4.2.3.5 Conclusion

This section has attempted to find at least partial answers to certain questions by examining the mapping of a metaphorical expression and linking this with an expression's chances of surviving the translation process. As will be the case in most of the sections that follow, the approach adopted involves both the use of a simple quantitative analysis and the application of human intuition in interpreting the results to derive qualitative conclusions.

Throughout this section, each example presented represents one of a potentially unlimited number of possible solutions that the translator(s) could have opted for, and we must therefore be careful not to overgeneralise any conclusions drawn. In addition, numerical data has to be interpreted carefully. It has consequently been necessary to drill down into the detail of the data in order to reach a clear picture of the precise processes that are involved.

Although there are some exceptions, it is not possible to report that translators appear to treat metaphorical mappings as high-level entities in their own right and make global decisions regarding how best to shift the metaphorical expressions associated with them into a particular target language. I therefore have to concur with Snell-Hornby's (1995:59–60) conclusion that the influence of the source text can frequently override a translator's feelings as to what is most in line with existing target-language metaphorical patterns.

Larger metaphorical structures such as mappings are therefore by and large preserved in translation, although there will be differences of detail between the extent and number of source and target text mappings. Metaphorical expressions tend to be dealt with on an individual basis, with some inconsistency as to how similar examples are handled. In terms of the question of whether mappings are shifted (i.e. translation shifts all tending in a particular direction) or scattered (i.e. their pointing in conflicting directions), the former result is associated with many one-example mappings, although there is little significance in this result, while the latter describes what happens with multiple larger mappings because of the frequently inconsistent handling of individual examples. What we do not regularly see, however, is any clear preference for a particular mapping on the part of translators; if different preferences for mappings do exist in a target language as a whole, they are not commonly reflected in the practice of the translators whose output has been analysed in this section. In general, the large number of omissions and replacements with non-metaphorical material is worthy of note, with results in some mappings being neither shifted nor scattered, but lost. Overall, there is little clear evidence of obvious manipulation, let alone subversion, of source-text metaphors. This is not to say that this does not take place, but perhaps this negative finding is not surprising given that the analysis has mainly taken place at the level of mapping (rather than individual expressions)—a level on which most translators do not seem to make conscious decisions. Similarly, the possibility that there may be some connection between mapping and procedure is tantalising but rather difficult to establish.

There is plenty of scope for further study of this aspect of metaphor in translation. An obvious extension would be to look at a number of thematically linked mappings in greater detail. The topic of mapping clusters could also be further investigated in this connection. Finally, more specific phenomena, such as culturally embedded mappings or mappings that are added in translation, could also be profitably examined with the aim of building up a more complete picture of this area. In all of these possible extensions, the main question would be whether it is possible to discover any regularities in the way that expressions belonging to a particular mapping are translated.

Notes

1. Since the context always makes it clear which language version is being referred to, I have decided to designate source-language and all target-language versions just by year of publication (rather than differentiating between them as 2004a, 2004b, etc.).
2. Section 4.2 is derived in part from an article published online in *The Translator* on 21 February 2014, available online at http://www.tandfonline.com/http://dx.doi.org/10.1080/13556509.2011.10799491 (Shuttleworth 2011).

Interlude One

Metaphors of Nature

The cluster of 6 mappings that relate to nature is attested in a total of 18 examples contained in the English data, 17 of which appear in 'Evolution Encoded', a situation that makes for a highly heterogeneous metaphorical discourse on the subject of nature. There are also four expressions—each representing a new mapping—that have been added in one or other language. As one might expect, most expressions are based on humanisation of the impersonal force of 'nature' (a concept that is itself a metaphor for the outworkings of both micro- and macroscopic physical 'laws'—another concept that is in turn also a metaphor . . .). However, in most examples, the general notion of nature being a person is fleshed out in a number of different images, some of which are highly specific in nature. The metaphorical expressions that relate to this cluster of mappings appear to be predominantly either discoursal or evaluative, with a small number having an exegetical function (see Section 5.1).

In a long and detailed historical discussion of the metaphors used to conceptualise nature, Norwick (2006) considers the plethora of tropes that have been employed in Western civilisation over the last 3,000 years and more. These include a large number of reifying metaphors (e.g. the Tree of Life, the Book of Life, the Clockwork Universe and the Web of Nature) and also some key personifications (e.g. nature as a mother, earth as a mother, nature as *Gaia* and nature as a handmaiden) that have played a central role in modelling the concept of nature as it has developed over the millennia.

Norwick's great strength lies in the breadth of his survey, which takes in the mythologies of many of the major cultures that have existed over the last three millennia across the territory that stretches from Western Europe to India. What he does not attempt, however, is any kind of analysis of the minutiae of text-based metaphors, offering instead a relatively general discussion of the topic. In this respect, his data is complemented by the data studied in this book. As we will see, as with the genetic material cluster, this cluster of humanising mappings appears to be a remarkably stable area in terms of retention of mappings, although, unlike this other cluster, some of the mappings and expressions contain a clear culture-specific element. The four most common mappings represented in the English data are as follows:

NATURE IS A CREATOR OF CODE (nine instances: e.g. ***Nature's code*** *minimizes the effects of genetic errors*: Freeland & Hurst 2004:87)
NATURE IS AN EXERCISER OF CHOICE (three instances: e.g. *Sophisticated computer searches can certainly improve on* ***nature's choice***: Freeland & Hurst 2004:90)
NATURE IS A BLIND DESIGNER (two instances: e.g. *Natural selection is a* ***'blind designer'***: Freeland & Hurst 2004:90)
NATURE IS A SOFTWARE DESIGNER (two instances: e.g. ***nature's skill as a primordial software designer***: Freeland & Hurst 2004:91)

To the extent to which one can claim that the overall effect of these mappings is to produce a coherent 'image' of nature, one can characterise their main entailments as presenting nature as a sentient, decision-making creator or designer. Although the purpose of the expressions is predominantly either evaluative or discoursal, I cannot resist making the observation that the overall picture of nature that emerges is more in line with the notion of 'intelligent design' than with the theory of evolution that is presumably subscribed to by all *Scientific American* authors—with the exception, of course, of the 'blind designer' expressions. In other words, in their efforts to communicate as clearly and graphically as possible, they have opted for images that are pre-scientific in character and thus, presumably unintentionally, have portrayed nature in a manner that flies in the face of their consciously held worldview.

While this is a relatively coherent image, the presence of other mappings—NATURE IS AN ARBITER and NATURE IS AN EXPERIENCER OF NEEDS, each accounting for one expression ('nature apparently *has not deemed* such a thing fit enough' (Gibbs 2004:75) and 'nature *needed* no figurative punctuation' (Freeland & Hurst 2004:87) respectively)—indicates that this area is not characterised by a totally homogenous metaphoricity. What is arguably the most significant mapping, NATURE IS A CREATOR OF CODE, will be looked at in some detail shortly. However, in terms of a more 'broad brush-stroke' impression, what happens to this image in the translated texts?

As may be expected, the picture is somewhat mixed. First of all, it is these two non-related mappings that are most vulnerable in translation: NATURE IS AN EXPERIENCER OF NEEDS vanishes in all translations and NATURE IS AN ARBITER in all but Russian and Italian. In contrast, of the mappings that are in line with this image of nature (each of which is in any case represented by more than a single expression), only NATURE IS AN EXERCISER OF CHOICE and NATURE IS A SOFTWARE DESIGNER fall by the wayside at all, in German and Polish, and French and Italian, respectively. So the central ones prevail, although all this might indicate is that they are central because they each have more instantiations.

Of the four individual mappings, NATURE IS A BLIND DESIGNER is retained almost universally, followed by NATURE IS A SOFTWARE DESIGNER and NATURE IS AN EXERCISER OF CHOICE. NATURE IS A CREATOR OF CODE is significantly depleted in a number of translations, although, on the other hand, some individual expressions undergo a certain intensification, as we will see presently.

Finally, the following new mappings are created in the production of the various translations, each represented by a single metaphorical expression:

German: NATURE IS A MECHANISM and NATURE IS AN IMPLEMENTER
Polish: NATURE IS A SOLVER OF PROBLEMS, NATURE IS A USER OF TOOLS and NATURE IS A USER OF CODE
French: NATURE IS SKILLED
Italian: NATURE IS A PLANNER
Chinese: NATURE IS A FILTERER and NATURE IS A THINKER

Most of these are quite clearly in line with the 'intelligent design' image cluster we have been discussing; however, the small number of metaphorical expressions involved makes their impact on the image of nature somewhat marginal.

So while it would be deeply satisfying for a researcher such as me to report that the image of nature was in some way overturned or transformed by at least some of the translators, the more mundane truth is that it is basically preserved, although the subtle patchwork created by the various relevant mappings and expressions is somewhat different in each case. In this instance at least, it appears that scientific translators are as open to unscientific language as the scientific authors themselves.

Turning now to look in more detail at the main mapping, it is interesting to observe what happens to the nine examples of NATURE IS A CREATOR OF CODE as in each target language, numbers of occurrences of this mapping are greatly depleted (the precise numbers of examples surviving in each target language being six in French, five in Chinese, three in German, Russian and Italian, and one in Polish). Almost exclusively, other instances have been removed (this happens 25 times), with six omissions and two instances of a new mapping. Two typical examples of this mapping are as follows:

Nature's code minimizes the effects of genetic errors
(Freeland & Hurst 2004:87)

At least 16 organisms . . . deviate from **nature's standard code**
(Freeland & Hurst 2004:91)

The possessive construction is not only found in all nine examples but is also vital for the existence of this mapping: convert it to another construction (*the natural code*, for example) and the (albeit weak) entailment that nature has created this genetic code vanishes. Indeed, this is precisely what we often see happening when such expressions are translated:

English: only 100 of one million alternatives had a lower error value than **nature's code** (Freeland & Hurst 2004:90)

Polish: tylko 100 kodów na milion ma niższy współczynnik błędu niż kod naturalny (Freeland & Hurst 2004:62)
[only 100 codes in a million have a lower level of error than the natural code]

Italian: solo 100 su un milione erano migliori di quello naturale (Freeland & Hurst 2004:94)
[only 100 out of a million were better than the natural one]

(These are just examples; in actual fact, this expression is removed in all target languages.) As discussed in Section 5.2.2, these examples constitute instances of grammaticalised metaphorical expressions. Interestingly, a quick check on Google indicates that direct equivalents of the phrase *nature's code* do not occur very frequently in most target languages. At the same time, this also seems to be the case for many of the alternative variants settled on by the translators. In other words, in this instance, the translators are avoiding close translations that in most cases are not commonly used in their target language, but using in their place ad hoc solutions that also do not enjoy a high degree of currency.

In all, there are arguably 12 examples of a direct equivalent of 'nature's . . . code' being used in any language (six in Chinese, four in French and one each in Italian and German, although in the case of the Chinese, the wording is neutral between a possessive and non-possessive reading and consequently a metaphorical and non-metaphorical interpretation) out of a theoretical maximum of 54. However, the grammatical restructuring illustrated earlier is not the only way in which this mapping is removed. In Russian, for example, 'nature's code' is paraphrased as follows:

Russian: только у 100 альтернативных кодов из миллиона цена ошибки меньше, чем у канонического (Freeland & Hurst 2004:90)
[only in 100 alternative codes out of a million is the price of an error lower than in the canonical one]

This same variant, *kanonichesky kod*, is used to render a second instance of this mapping (Freeland & Hurst 2004:60), and on a third occasion a similar solution, *genetichesky kod* ('the genetic code': Freeland & Hurst 2004:55), is employed by the translator.

The few examples of the Russian expression *kanonichesky kod* that can be located via Google appear to refer to something other than genetic information. This could possibly lead us to conclude that translators sometimes supply a formulation of their own that has not hitherto been associated with that context in the target language, but on other occasions simply opt for expressions that are already an accepted way of talking about the object in question.

Along similar lines, the Italian translator uses the expressions *presente in natura* ('present in nature': Freeland & Hurst 2004:94) and *esistente in natura* ('existing in nature') on a total of three occasions (Freeland & Hurst 2004:91, 95). Similarly, 'nature's real code' (Freeland & Hurst 2004:86) is translated into Chinese as 遺傳密碼 ('heredity code' or 'genetic code': Freeland & Hurst 2004:78).

While phrases based on the 'nature's . . . code' pattern rarely receive a direct translation, within the examples of this mapping that survive in the target languages, there is a strong tendency towards intensification, as can be seen in the following examples:

English: many came to view **nature's real code** as little more than a random accident of history (Freeland & Hurst 2004:86)

German: . . . erschien **das von der Natur realisierte System** daher vielen Wissenschaftlern als wahllos zusammengestückelt—das Ergebnis blinden Zufalls (Freeland & Hurst 2004:89)
[. . . to many scientists **the system realised by nature** thus appeared as randomly cobbled together—the result of blind chance]

French: on en vint à considérer **le code élaboré par la nature** comme un aléa de l'histoire (Freeland & Hurst 2004:62)
[people came to consider **the code elaborated by nature** as a chance of history]

English: **Nature's code** minimizes the effects of genetic errors (Freeland & Hurst 2004:87)

Russian: **Код, созданный природой,** сводит к минимуму последствия ошибок (Freeland & Hurst 2004:58)
[**The code created by nature** reduces the consequences of errors to a minimum]

Polish: **Stworzony przez przyrodę kod** minimalizuje efekty błędów genetycznych (Freeland & Hurst 2004:60)
[**The code created by nature** minimizes the effects of genetic errors]

The intensification that can be observed in these examples should be seen against the fact that six out of the nine expressions motivated by this mapping are removed or omitted in the German, Russian, Polish and Italian translations, and three out of the nine in the French and Chinese. In general terms, the intensification that we observe is accompanied in each case by explicitation of the mapping's implicit entailment. These examples, along with a handful of other similar instances, indicate that the translators may opt for bolder, more definite instantiations of the mapping in preference to more direct renderings. (The existence of these metaphorical renderings also demonstrates that the identification of the original English expressions as (marginally) metaphorical is not fanciful.)

There are instances where this intensification is accompanied by a change in mapping. The overall effect of both of these kinds of decision on the part of translators is to propagate, or at least accentuate, source-text expressions such as *nature's code* that may only be marginally—or even unintentionally—metaphorical, even though in many cases the expression chosen by the translator is not a standard target-language phrase. This does of course occur alongside the alternative, stronger tendencies that have also already been commented on, of moving away from this mapping and/or of intensifying the metaphor. Furthermore, certain source-text wordings appear to trigger these last two tendencies very readily across multiple languages:

English: **Nature's version** [of the genetic code] looked less elegant than several of the theorists' hypotheses. (Freeland & Hurst 2004:86)

German: **die Lösung, welche die Natur gefunden hatte,** erschien weniger elegant als einige theoretische Vorschläge. (Freeland & Hurst 2004:87)
[**the solution found by nature** appeared less elegant than a number of theoretical proposals.]

Russian: **То, что придумала природа,** выглядело гораздо менее изобретательным, чем многие гипотезы ученых. (Freeland & Hurst 2004:55)
[**What nature thought up** seemed much less inventive than many scientists' hypotheses.]

Polish: **Wariant, którym posłużyła się przyroda,** okazał się znacznie mniej elegancki niż niejedna z hipotez. (Freeland & Hurst 2004:58)
[**The version that nature used** was significantly less elegant than a number of hypotheses.]

French: **la méthode choisie par la nature** était moins élégante que certaines des hypothèses qui avaient été proposées (Freeland & Hurst 2004:60)
[**the method chosen by nature** was less elegant than certain of the hypotheses that had been proposed]

Italian: **La versione elaborata dalla natura** sembrava meno elegante di molte delle ipotesi formulate dai teorici. (Freeland & Hurst 2004:90)
[**The version elaborated by nature** seemed less elegant than many of the hypotheses formulated by the theorists.]

It is not easy to identify the most likely reason why this particular expression should evoke such a remarkable correspondence of response on the part of five out of the six translators, except to suggest perhaps that this expression happens to be discoursal and so more prone to frequent modification (see Section 5.1). Only in Chinese do we see a more direct translation, as follows:

Chinese: 跟科學家提出的幾種假說比較起來,**自然界的版本**實在不怎麼漂亮。(Freeland & Hurst 2004:78)
[compared to several hypotheses proposed by scientists **nature's version** is really not particularly beautiful.]

One possible hypothesis—that the translators opt for the solutions they do because the word *code* is absent from the expression—turns out to be incorrect, as an expression containing the phrase *nature's own* is universally removed or omitted. Quite clearly, there appears to be something occurring with the translation of this expression that it is not currently possible to explain.

Taking these facts together, what we have seen through the other examples is a tendency to lose some mappings through apparently unmotivated rewording, but also some instances of individual expressions being intensified through explicitation. Overall, however, the picture of nature as an agent of intelligent design remains basically intact, if different in detail, in all target languages.

5 Intuitive Classifications of Metaphor

I am now going to turn my attention to two parameters that are not inspired by any particular theoretical framework. These are the purpose for which a metaphorical expression seems to be used and the level of conventionality by which it is characterised. The term 'intuitive' is designed to allude to the relatively informal, non-theoretical nature of these two parameters.

5.1 Purpose

It is perhaps the parameter of purpose that is one of the most central to the scientific nature of the texts that are being studied: as will be explained in the following section, three of the purpose types—exegetical, terminological and theory-constitutive—can be thought of as relatively central to the purpose of communicating scientific ideas, while the other two—discoursal and evaluative—are more peripheral to this purpose, as they are more involved with engaging a non-specialist readership than with the more central scientific processes of concept formation and explanation. The expectation is of course that the former will perhaps be less negotiable than the latter in terms of the range of procedures that are most frequently employed by the translators.

This chapter section follows broadly the same structure as most of the others, with the important difference that a large proportion of the practical part focuses on two extended examples of metaphorical terminology in translation.

5.1.1 Background

All in all, people choose to express themselves through metaphors for a number of different reasons. These would include both clarifying and concealing the writer's or speaker's intended meaning, and presenting fresh and original insights in a striking or memorable fashion. In addition, metaphorical modes of expression are often unthinkingly adopted simply because much-used mappings have become embedded in the language that a person speaks. In terms of metaphor in translation, it seems intuitively at least reasonable that the purpose for which a metaphorical expression is included in the source text—whether, for example, it seems to be of central or peripheral importance—might influence the amount of attention that translators will pay to it, and hence the possible procedures that might be used to translate it.

Boyd famously distinguishes two main types of scientific metaphor, which he terms 'exegetical' (or 'pedagogical') and 'theory-constitutive'. The former plays a role in the teaching or explaining of theories that 'already admit of entirely adequate nonmetaphorical (or, at any rate, less metaphorical) formulations' (1993:485); in other words, they do not form an essential part of any scientific theory but are merely employed by a speaker or writer as a means of making a particular scientific concept clearer. Theory-constitutive metaphors, on the other hand, represent an 'irreplaceable part of the linguistic machinery of a scientific theory: cases in which there are metaphors which scientists use in expressing theoretical claims for which no adequate literal paraphrase is known' (1993:486). These may only exist on a temporary basis while the language used to frame a particular concept is in the process of stabilising. Examples of these and other kinds are provided below.

In her discourse-based work on metaphor in architectural reviews, Caballero distinguishes the three roles of metaphor in the genre in question as description, evaluation and authorial commentary (2003b:158–64). The first of these is relatively specific to the architectural genre that she discusses, and since it is largely reflected in image metaphors, it does not specifically concern us here. The evaluative role is linked with the rhetorical goal and structure of the texts. Finally, she sees the last role, that of authorial commentary, as being filled in expressions that frame buildings as 'the agents of mental processes' (2003b:163). An example of this would be 'As Zapata's building *demonstrates*' (2003b; emphasis original), and she interprets the inclusion of such expressions as 'a useful device for covert authorial comment' (2003b:164). This category is located in the context of Low's work on personification in academic texts (1999).

5.1.2 How These Concepts Feature in the Data

As discussed in Section 5.1.1, metaphor can have one (or more) of a number of different purposes. Purposes will clearly often be mixed, but for the sake of simplicity, I have classified each expression according to what I consider to be its principle one. As will have been gathered by now, the study makes use of five categories, which are discussed in the following sections.

5.1.2.1 Discoursal

This type is used as a device for permitting a limited comment to be made and also frequently as a technique for avoiding a more periphrastic wording. Discoursal metaphor is largely but not exclusively associated with humanisation and/or personification (e.g. 'amino acids *like* to cluster together', as opposed to 'amino acids *have a tendency* to cluster together', for example). Besides this, many metaphorical expressions that are the single instantiation of a particular mapping (e.g. 'ATF is *neatly packaged* inside the tips of axons', Fields 2004:57, which is the only expression associated with the mapping that I have verbalised as A CONTAINER IS A PACKAGE) belong to this type. Many 'general-purpose' scientific metaphors (e.g. 'Einstein taught us the value of daring to *think outside the box*', Fields 2004:61) that occur in all kinds of scientific discourse are also included in this category. Most but not all examples fall within Caballero's concept of 'authorial commentary' (2003b:158–64) that was discussed in Section 5.1.1. Along with those that have an evaluative purpose, some expressions that are discoursal in nature belong to the metadiscourse of the article in which they appear.

Metaphorical expressions in this category represent a kind of scientific shorthand, or at times are simply examples of catachresis (or, in other words, the process of forming a new metaphorical usage in order to give a name to something that was previously nameless: see, for example, Black 1962:33). The first of these two types, at least, is also likely to be much more closely associated with popular rather than professional science, which could mean that, along with evaluative expressions, owing to their peripheral status, they are ultimately more dispensable and may well survive less readily in translation.

Half the small number of added expressions are discoursal.

5.1.2.2 Evaluative

Metaphorical expressions of this type involve wordings that seem to have been selected for their connotations or clear culture-specific meaning (e.g. 'if glia could *chatter*', Fields 2004:56; 'tiny cellular *butlers*', Freeland & Hurst 2004:86). Many of these derive from one-off mappings and, as can be seen from the examples just quoted, can give rise to rich images. Like discoursal expressions, examples in this category are also thought of as peripheral to a text's main scientific purpose.

5.1.2.3 Exegetical

This is the first of the three categories that I classify as 'central' to scientific purpose. Expressions with an explanatory function, whether formal ('*cognitive control mechanisms*', Solms 2004:87; '*synonymous* codons', Freeland & Hurst 2004:87) or informal ('neurons as the brain's *prime communicators*', Fields 2004:55; 'synaptic *firing*', Fields 2004:56; 'think of it as *Life, version 2.0*', Gibbs 2004:76)—the latter being far more common in my data than the former—are termed 'exegetical'. Also included in this category are expressions that are descriptive in nature (cf. Caballero 2003b:158–64); these are based on physical attributes (e.g. 'additional *buds*, termed dendritic spines', Nestler & Malenka 2004:83; 'oligodendrocyte glia cells *wrap* around axons', Fields 2004:56; 'Schwann glia cells had formed a *bridge* across the divide', Fields 2004:61). These expressions are in fact also image metaphors (even though some image metaphors are categorised as 'evaluative'—e.g. 'unneeded synapses *get pruned*', Fields 2004:61).

5.1.2.4 Terminological

This is the second category that I consider to be 'central'. Much scientific terminology, whether old and established (e.g. *disorder*, Javitt & Coyle 2004:50; *dendrite*, Fields 2004:55) or recently coined (e.g. *neurotransmitter*, Fields 2004:55; *BioBrick*, Gibbs 2004:76), is created on the basis of metaphor, although in some languages the metaphor is concealed in the etymology of the term, as discussed in Section 1.2.3.

In line with this, some terms appear to be less formal than others (e.g. *wetware*, Gibbs 2004:78), while as I hope to show there is some evidence that translators do not always treat items of terminology as isolated lexical units but do sometimes pay some attention to the surrounding discourse when deciding on the best solution. Another way in which translators occasionally intervene is to engage in secondary term formation, a process originally defined by Sager as the practice of creating a term in one language for a known concept for which a term already exists in another (1990:80).

5.1.2.5 Theory-Constitutive

The third 'central' purpose type, as commented by Boyd (see Section 5.1.1), is the kind of scientific metaphor used when no suitable literal alternative is available (e.g. 'how genes *encode* proteins', Freeland & Hurst 2004:84, 'when a DNA gene is *expressed*', Freeland & Hurst 2004:87). Furthermore, as discussed in Section 1.2.2, theory-constitutive metaphor can sometimes guide the development of scientific ideas (e.g. '*assemble* [genetic] systems *from pieces*', Gibbs 2004:76; 'creating . . . biological *devices*', Gibbs 2004:75; '*libraries* of interchangeable DNA parts', Gibbs 2004:75).

As can be seen from these descriptions, along with the mapping parameter, it is here that the categories are most closely linked to the scientific subject matter of the texts in my data. Clearly, the functions I have proposed overlap to a large degree, and it is also possible for a metaphorical expression to serve more than one function, so in many cases, it is only practicable to suggest which appears to be the predominant one; for example, an expression such as '*Like a wave of cheering fans sweeping across a stadium*' (Fields 2004:60) seems to possess both exegetical and evaluative force, although it is categorised as exegetical because that seems to be the main function that it serves in the context in which it appears.

5.1.3 Translation Procedures and Analysis of Examples

This section starts with the usual look at the numerical background. However, as stated earlier, it then focuses to a very large extent on one particular purpose type with a detailed examination of how two items of terminology are translated into the various target languages.

5.1.3.1 Numerical Background

In this section, in an attempt to highlight what happens to the purpose categories as a whole when translated, we will be looking at how the numbers of examples of each category in the source texts contrast with those in the target texts. The figures that we will be examining are contained in Table 5.1:

Table 5.1 A Comparison of the Numbers of Metaphorical Expressions in Each Purpose Category in the Source and Target Texts

Source-language purpose categories (numbers of source-language expressions shown in brackets)	*Target-language purpose categories (all six target languages taken together)*				
	Discoursal	*Evaluative*	*Exegetical*	*Terminology*	*Theory-Constitutive*
Discoursal (68)	207	6	4	0	0
Evaluative (66)	16	142	9	3	0
Exegetical (101)	1	6	385	2	5
Terminology (18)	0	0	3	62	0
Theory-Constitutive (15)	0	0	0	1	53

The main observation to make is that overall there is very little shifting between different purpose categories in translation: for each source-language category, the great majority of expressions remain in the same category in translation, with only a small number of expressions being shifted into a different category. However, this is perhaps unremarkable, given the nature of the texts; after all, it is unlikely, for example, that large numbers of discoursal expressions will take on a more exegetical function as a result of translation.

There are some qualifications that need to be made to this general observation, but in the greater number of these cases, the shifts are between two categories that are each either central or peripheral. Thus, for example, a relatively significant number of evaluative expressions are replaced by discoursal expressions, with a smaller number even being replaced by exegetical ones. On the other hand, we do not see anything like even such a modest level of shifting in the case of discoursal expressions, although a small number are converted to examples of the evaluative and exegetical categories. Taken together, shifts between these three values account for most of the total number of shifts. The exegetical, terminology and theory-constitutive categories, on the other hand, are the best preserved.

A glance at the language-specific data indicates that no one target language skews this data to any significant extent, even though the Chinese translators, for example, produce hardly any such shifts at all.

To close this section, the following observations are worth making:

i) The proportionality between the different purpose categories remains stable in translation.
ii) There are some shifts between pairs of categories both of which are central or peripheral (e.g. evaluative to discoursal).
iii) On the other hand, there are far fewer shifts between central and peripheral categories (but a small number from evaluative to exegetical).
iv) There are also relatively fewer examples of metaphorical expressions shifting from the discoursal purpose category.
v) Exegetical, terminology and theory-constitutive—the three central categories—are the best preserved in translation.

The matters raised by these points will be taken up at least briefly in the next section.

5.1.3.2 *Examples of the Different Procedures*

Overall, there is relatively little shifting between different purpose categories in translation (perhaps unsurprisingly, given the nature of the texts). Each source-language category is predominantly preserved in translation, with expressions that are shifted into another category being very much in the minority.

If most of the following sections are a bit shorter than they might have been, this is in anticipation of the detailed analyses offered in Chapter Six, where many procedures that are not specific to a particular parameter are brought to the surface.

5.1.3.2.1 DISCOURSAL

As pointed out earlier, evaluative and discoursal metaphorical expressions appear to be the least stable in translation, although this tendency is less pronounced in the case of discoursal expressions. Plenty of such expressions are removed or omitted, although only a very small number are shifted into evaluative and exegetical expressions.

The following will serve as an example of the former:

English:	local **conversations** are necessary . . . neurons and glia carried on more discrete conversations (Fields 2004:60)
German:	Dass ein lokales **Geflüster** zwischen Glia- und Nervenzellen existiert (Fields 2004:54) [That a local **whispering** exists between glia and nerve cells]

Here an essentially neutral word ('conversations') is replaced by a more figurative alternative ('Geflüster') that means 'whispering'.

In the next example, the general-purpose item 'pathway' is substituted in the French translation for the more specific 'circuit', which is in line with the mapping THE NERVOUS SYSTEM IS CIRCUITRY that is predominantly associated with exegetical expressions.

English:	various drugs of abuse ultimately lead to addiction through a common **pathway** (Nestler & Malenka 2004:80)
French:	la dépendance provoquée par divers stupéfiants dépend d'un même **circuit** cérébral. (Nestler & Malenka 2004:42) [the dependence caused by various drugs depends on the same brain circuit]

This translation is echoed in another, similar expression, in which *neuronal* ***pathway*** (Solms 2004:85) is rendered as *un* ***circuit*** *neuronal* (Solms 2004:78, 'a neural **circuit**'), in spite of the existence of the more literal alternative *voie neuronale* ('neural pathway').

5.1.3.2.2 EVALUATIVE

Exceptions to the relative stability of this category that was noted earlier include a comparatively small number of expressions becoming discoursal, rather fewer becoming exegetical and a marked tendency towards removal or omission. Overall, however, retention remains the commonest procedure.

The following example illustrates how an evaluative expression can be shifted into discoursal ones:

English:	nature apparently **has not deemed** such a thing fit enough to survive in the wild. (Gibbs 2004:75)
Russian:	но **природа решила,** что такая конструкция будет недостаточно жизнестойкой. (Gibbs 2004:47) [but **nature decided** that such a construction would not be strong enough to withstand life.]

Polish: **Przyroda**, niestety, **nie uznała** tego za istotne i w toku ewolucji taki licznik nie powstał. (Gibbs 2004:59)
[**Nature**, unfortunately, **did not recognise** this as important and in the course of evolution no such counter came into being.]

Chinese: 顯然自然界**並不認為**這種細胞有足夠的適應力可以存活下來。(Gibbs 2004:87)
[the natural world **really did not consider** that this kind of cell had sufficient power of adaptation to have been able to survive.]

The difference between the evaluative English expression and the discoursal versions found in these translations is that, while all involve an unscientific humanisation of nature, in the English expression, the word 'deem' smacks more of the making of arbitrary decisions—an overtone that is not present in the translations.

Along similar lines, some evaluative expressions become exegetical, such as the following:

English: if glia could **chatter** (Fields 2004:56)

French: si les cellules gliales **communiquent** (Fields 2004:57)
[if the glial cells **communicate**]

Italian: se fossero state in grado di **comunicare** (Fields 2004:50)
[if they had been capable of **communicating**]

Chinese: 神經膠細胞如果可以**對話** (Fields 2004:36)
[if glial cells could **dialogue**]

Here the idea of cells communicating or dialoguing is a statement of a scientific idea, while talk of 'chattering' adds a clear evaluative colouring.

Although clearly not specific to this kind of expression, quite a number of evaluative metaphors are removed in different target languages, including this more unusual case that involves a proper name:

English: **a 'Goldilocks' genetic circuit, one that lights up when a target chemical is present but only when the concentration is not too high and not too low** (Gibbs 2004:78)

Russian: Goldilocks—устройство, включающееся только при строго определенной концентрации химического сигнального вещества. (Gibbs 2004:50)
[Goldilocks, a device that only switches on at a strictly defined concentration of a chemical signalling substance.]

Polish: genetyczny obwód o nazwie Goldilocks. Powoduje on, że komórki świecą w obecności konkretnego związku chemicznego, przy czym dzieje się tak tylko wtedy, kiedy jego stężenie ma odpowiednią wartość (Gibbs 2004:62)

> [a genetic circuit called Goldilocks. It causes cells to light up in the presence of a concrete chemical compound, although it only operates when its concentration has the right value]

Although the name itself is retained in these translations, the absence of an explanation is likely to make the metaphorical sense unavailable to a large proportion of the readership.

5.1.3.2.3 EXEGETICAL

As noted earlier, limited numbers of the 101 English exegetical expressions are converted to either theory-constitutive or evaluative in translation; overall, however, instances of repurposing are very few and far between—unsurprisingly, of course, given the scientific centrality of this category of expression. While this represents a significant level of stability, it should also be borne in mind that exegetical expressions are removed or omitted on around 30% of all occasions.

Of course, the set of exegetical data is not just a collection of isolated expressions, as many of the examples contribute to one of a number of thematic clusters that relate to topics such as the circuitry of the brain, the encoding of genetic information in 'linguistic' units, the 'firing' of neurons, the tree-like physical shape of brain cells and the various roles attributed to nature in the act of creation. The mapping THE NERVOUS SYSTEM IS CIRCUITRY is represented by 12 expressions, 8 of which consist of the technical noun *circuit* (or, in one case, *circuitry*)—with or without a modifier such as *neuronal*, *complex* or *genetic*—and 4 of which are more complex and/or informal ('neuronal *circuits* that are not *wired together* directly': Fields 2004:61) or in some way conceptually obscure to a non-scientist (*reward circuitry*). Given the multilingual nature of the data, this gives rise to a large number of (admittedly interlinked) translation examples. With few exceptions, translators seem to retain the metaphoricity of the simpler examples while removing or omitting most of the more complex ones.

Examples of the latter tendency include 'neuronal *circuits* that are not *wired together* directly' (Fields 2004:61), which receives a metaphorical translation in none of the six target languages, and 'Drug abuse produces long-term changes in the *reward circuitry* of the brain' (Nestler & Malenka 2004:78), the metaphoricity of which is omitted in three out of the five renderings. As on other occasions where this happens, it is unclear why this first example should lose its metaphoricity so universally.

On the other hand, a counter-example is supplied by the expression 'The mental picture most people have of our nervous system resembles a *tangle of wires that connect neurons*' (Fields 2004:55), which is retained in all target texts (even though it is once again unclear why this should be the case). Quite clearly, having one out of four examples unequivocally contradicting what is suggested by the other three is a serious drawback, although on the other hand it does appear that the contention that the simpler examples tend to be retained may still be able to stand.

In the case of expressions that instantiate the NEURONS ARE TREES mapping—which have already been examined in some detail in Section 4.2.3.2.2—it is not possible to discern a single clear factor that determines how a particular expression—of those eight expressions that fall within the exegetical category—will be translated. The following example is typical of the wide, almost random diversity of approaches that can be observed among these examples:

English:	Each neuron has **a long, outstretched branch** (Fields 2004:55)
German:	[omitted]
Russian:	Каждый нейрон снабжен **одним длинным отростком** (Fields 2004:24) [Each neuron is equipped with **a single long shoot**]
Polish:	Od każdego neuronu odchodzi **długa wypustka, najczęściej rozgałęziona na końcu** (Fields 2004:28) [From each neuron there emerges **a long projection, usually branching at the end**]
French:	Chaque neurone a <u>un long prolongement</u> (Fields 2004:57) [Each neuron has <u>a long extension</u>]
Italian:	Ogni neurone ha **una lunga ramificazione molto estesa** (Fields 2004:50) [Each neuron has **a long, very extended ramification**]
Chinese:	每一個神經元有**一條往外延伸、長長的分支** (Fields 2004:36) [Each neuron has **an outward extension, a long branch**]

Here we have fairly straight retention (Chinese), retention but in a more abstract form (Russian, Polish and Italian), removal (French) and omission (German).

5.1.3.2.4 TERMINOLOGY

Terminology has been selected for a much more detailed discussion because a) relatively little has been written on this important subject so far (although, actually, much the same may be said of the other categories too) and b) the degree of mediation that takes place in the case of these apparently stable units defies expectation. It is perhaps striking that the data only contains 18 examples of this purpose category. However, this could be a result of the fact that specialist terminology is often avoided in popular science, as discussed by Olohan (2016:188; see Section 1.2.4). In addition, it is likely that a significant number of dead metaphors would also be included in this category, not all of which have been recorded, in line with the policy outlined in Section 1.2.3. Even though this is the case, it is not entirely clear—and this is beyond the scope of this research—what the proportion of metaphorical terminology is in scientific language as a whole.

Apart from the work done by Temmerman (2002, 2010), as far as I am aware, the next few pages contain one of the first studies on metaphorical terminology in translation. Along with theory-constitutive metaphorical expressions, it is metaphorical terminology that seems to be best preserved in translation, although some modifications, removals and omissions are also encountered. The 18 English expressions that were classified as terminological yield a total of 96 translation examples and include instances of both core terminology and terminological collocations. All of these were looked at in the process of building up the following picture: on the basis of the small amount of data available, translators appear to pay attention to the following kinds of factor, even when translating a metaphorical expression as relatively stable as a terminological item.

The major observation is that in most or all cases, the target-language equivalents seem to have already been coined by the time they feature in translated versions of *Scientific American*, with the result that there appear to be few examples of the process of secondary term formation contained in the data. This means that translating many terms entails using an accepted equivalent, as that is simply the most obvious solution. In practical terms, use of this corresponding item can entail 'straight' translation, borrowing, complete or partial calquing, explicitation, approximation, simplification or reduction, for example, depending on the precise nature of the target-language item. That said, locating and inserting the correct terminological equivalent is not the only translation issue involved, as will be seen in the two extended examples that follow.

These two practical examples of metaphorical terminology in translation are worth examining in some detail, as they appear to illustrate the kind of decision-making that translators will go through when presented with terminology that is likely to be novel or unfamiliar. The first example involves the English term 'wetware':

English:	Life runs on **'wetware'** (Gibbs 2004:78) (Earliest usage of 'wetware' in English-language pages indexed by Google: 1984)
German:	. . . **das nasse Element.** Mit »**Wetware**« zu arbeiten bedeutet . . . (Gibbs 2004:72) [. . . **the wet element.** Working with **'wetware'** means . . .] (Earliest usage of 'wetware' in German-language pages indexed by Google: 1996)
Russian:	Жизнь протекает <u>в жидкой среде</u> (Gibbs 2004:50) [Life proceeds <u>in a liquid medium</u>] (Earliest usage of 'wetware' in Russian-language pages indexed by Google: 1998)

Polish: Życie związane jest ze środowiskiem roztworów wodnych (Gibbs 2004:62)
[Life is linked with an environment of aqueous solutions]
(Earliest usage of 'wetware' in Polish-language pages indexed by Google: 1999)

Italian: [Entire English phrase omitted]
(Earliest usage of 'wetware' in Italian-language pages indexed by Google: 1996)

Chinese: 生命的運轉是在有水的環境中進行的 (Gibbs 2004:90)
[The operation of life is conducted within an environment that contains water]
(Earliest usage of 'wetware' in Chinese-language pages indexed by Google: 1997)

This term—which in this context refers to the human mind and the central nervous system as the infrastructure that an organism needs in order to function, and which is used in a sense analogous to hardware and software—occurs only once in the article 'Synthetic Life'. The term is not frequently used in scientific discourse; it is also found in science fiction, which is one of the factors that make its use somewhat evaluative. It seems that the original English term tends to be used in all the European target languages (this appearing to be particularly the case in Italian), while the calqued term 濕件 ('wet ware') has existed in Chinese since at least the late 1990s. The fact that the original English term started to occur in pages indexed by Google in all relevant target languages between five and eight years before the original article was published indicates that it was in principle available for translators to use. Yet as we see, the term is only retained in one target language (German) and here only with an accompanying paraphrase. The Russian, Polish and Chinese versions, on the other hand, rely solely on paraphrase, while the Italian omits any mention of the original phrase. In this particular instance at least, an innovative metaphorical term has been translated in a manner that has entailed some deliberate mediation—possibly because of a reluctance by the translators to retain too much English-language material—rather than by using a pre-existing equivalent.

Unlike 'wetware', the term 'BioBrick' is used a total of eight times in this same article, in which it refers to a key concept. Also unlike the previous term, BioBrick is theory-constitutive in nature, as it suggests the idea that there are free-standing, prefabricated genetic 'components' consisting of sequences of DNA, which can be combined with other similar elements to build a biological 'machine' to fulfil a particular function within a cell in a manner that is analogous to the way that mechanical parts are combined to create an 'actual' machine. (Note the fact that I chose the word 'build' here rather than 'create', as it is more congruent with the mapping that is being suggested.) The BioBrick metaphor is discussed in Hellsten and Nerlich (2011); within the present work, the term is of interest, as its translation probably involves some conscious decision-making, unlike many of the others where the translation is more automatic.

The following are five of the eight instances of this metaphor:

> Building with BioBricks.
>
> (Gibbs 2004:76)

> At a mechanical level, individual BioBricks (as the M.I.T. group calls the parts) can be fabricated and stored separately, then later stitched together to form larger bits of DNA.
>
> (Gibbs 2004:77)

> One of the vials in Endy's box, for example, contains an inverter BioBrick (also called a NOT operator).
>
> (Gibbs 2004:77)

> These new BioBricks were then added to M.I.T.'s Registry of Standard Biological Parts.
>
> (Gibbs 2004:77–8)

> But in the intervening year the inventory of BioBricks has grown, the speed of DNA synthesis has shot up, and the engineers have gained experience assembling genetic circuits.
>
> (Gibbs 2004:78)

What might be immediately apparent is that, although the central entailment of 'BioBrick' is that of construction, the term is in fact also used in contexts that are at odds with this, such as the production of fabric (see the phrase 'stitched together' in the second example) or the formation of logic circuits ('an inverter BioBrick' in the third). This hybridity will be discussed in detail below.

Following research that was being conducted at MIT, mentions of BioBricks started to appear on the Internet in about 2002, and for some years, the term has been the subject of a trademark. In the original *Scientific American* article, which dates from 2004, the term first appears twice on the same page, once in a photo caption and once in a subheading ('Building with BioBricks'). Given its repeated use in the article, rather than examining multiple source text—target text pairs complete with the specific decision that each translator took, on this occasion I am going to try to understand the translators' overall translation approach in a global manner as represented by the complete set of instances. Indeed, in the case of some of the target texts, there is in fact no one-to-one correspondence of instances of this term, as will be made clear in this section.

There seem to be at least four largely interlinking points that are worth making about the way that this metaphorical term is translated, as set out in what follows.

1. Interestingly, the number of instances of this term varies widely from one target language to the next, as follows:

German:	13 examples of 'BioBrick'/'BioBricks'
Russian:	6 of 'BioBrick'/'BioBricks'
Polish:	8 of 'biocegiełki' (partially calqued on BioBrick) and two of 'BioBricks'
Italian:	7 of 'biomattoni' (partially calqued on BioBrick) and seven of 'BioBrick'
Chinese:	8 of '生物磚' (calqued on BioBrick and incorporating 磚, the character for 'brick')

   (The article is not translated into French.)

   On the face of it, this wide variation strikes one as surprising, given that this is a piece of scientific terminology. The 13 examples in the German target text are explained by the policy of paraphrasing that seems to be pursued by *Spektrum der Wissenschaft*, which has already been mentioned. In the case of the Italian, the term 'BioBrick' is used consistently to translate the English term, while 'biomattoni' sometimes accompanies it, being used to translate a neighbouring co-referential item such as 'part' (see the listings under point 4.). 'Biomattoni' has also been added in the article deck and is used in the subheading 'Biomattoni da costruzione' ('Building with BioBricks').

2. Following on from this, the simultaneous use of two synonymous terms in Italian and Polish is worthy of note. As stated earlier, *Le Scienze* contains a more or less balanced use of 'BioBrick' and 'biomattoni', with the former possibly functioning as the standard equivalent but the latter appearing in the two prominent positions just described. This does not necessarily indicate that there may have been some uncertainty as to which was the preferable term and is more likely to be evidence of an attempt to explain the meaning or, more specifically, to make the rich-image basis of the term available to the readers of the Italian publication. The two uses of the term 'BioBricks' in the *Świat nauki* text, on the other hand, seem simply to serve the purpose of giving readers access to the MIT group's original designation.

   Ultimately, BioBrick appears to have become the accepted term in all five European target languages; however, at the time when the translations were produced, this was clearly far from being resolved, and the term was still in the process of being formed in at least some of the target languages studied. The question of whether any of the calques listed earlier were invented by the translator will be examined shortly.

3. Once again, following on from the previous point, the degree of explanation and meaning positioning that can be seen in each target text is also of interest. As we have already seen, some equivalents permit the 'brick' image to continue to be available in translation, and this can be seen to varying extents in the different target texts. No obvious explanation of the term's meaning is offered to the readers of *V mire nauki*, not least because the Russian word for 'brick' is nowhere present. Readers of the Italian, Polish and Chinese editions do at least encounter a transparent term, as described earlier; however, Chinese readers are provided with no indication of the original term (even though other English items, such as DNA and TNT, as well as the name of key scientist Drew Endy, are included in the text). The German translation, while it offers no alternative transparent form, is the only one to explain the origin of the term: 'Der Name [BioBricks] spielt auf das englische Wort für Ziegel- oder Bausteine an.' ['The name alludes to the English word for brick or building block'] (Gibbs 2004:70).

   At the same time, the amount of signalling in the form of inverted commas varies greatly among the texts. The German translation matches the English source text by only placing the first occurrence of the term in quotation marks. The Italian translator uses them for five of the seven occurrences of 'Biomattoni', but for only one instance of 'BioBrick'. This suggests that the former may well have been coined by the translator, an impression that is confirmed by a time-restricted Google Advanced Search. We see something similar in *Kexueren*: three out of the eight instances of 生物磚 are in inverted commas, and a Google Advanced Search indicates no relevant uses of the term predating the publication of the translation. On the other hand, while the same kind of search on Polish 'biocegiełki' suggests that this too is a translator coining, no inverted commas are used for any of its occurrences. None of the occurrences of 'BioBrick(s)' in the Russian text are placed in inverted commas (the fact that they all appear in italics like all other foreign expressions is not relevant).

   Finally, each version of the article introduces and positions the new term (or terms) in its own often complex manner. For example, as mentioned earlier, the Italian translator brings the Italian term to the readers' attention right at the beginning of the article and then interweaves instances of the two terms throughout the remainder of the text. At the fourth use of 'biomattoni', they are explicitly equated with the BioBricks of the MIT group ('i singoli «biomattoni»—o BioBrick' ['the individual "biomattoni"—or BioBricks']: Gibbs 2004:80). Next, the Polish, Italian and Chinese target texts include a close rendering of the explanatory 'As the MIT group calls them', while the Russian substitutes 'Endy' for 'MIT', and the German explains the origin of the term as mentioned earlier. Lastly, the term is arguably introduced with greater subtlety in Russian than in many of the other texts through the use of a carefully selected set of contextual co-referents that sees BioBricks being described using the words 'деталь' ('part'), 'элемент' ('element') and 'фрагмент ДНК' ('DNA fragment') in rapid succession (Gibbs 2004:48). On the same page, the translator opts for the image 'Биологический конструктор' ('Biological construction set') instead of a direct translation of 'Building with BioBricks'—a new rich image that may well have been inspired by Lego-type toys.

This last factor aside, bearing in mind the degree of English knowledge in the countries in question (see Section 2.1), it seems that the rich-image import of this metaphorical term—and hence its theory-constitutive force—is least accessible to readers of *V mire nauki*, and most accessible to the audience addressed by *Spektrum der Wissenschaft* or *Le Scienze*.

4. Again, building on the previous point, it appears that the collocates and contextual co-referents encountered in each of the texts may not all be homogenously metaphorical to the same degree, which suggests that the congruence of the surrounding text to a key metaphorical term may well be less than total. There is clearly wide scope for a writer's or translator's supporting lexical choices to either bolster or negate a metaphorical meaning, as I noted briefly earlier with respect to my selection of the word 'build' instead of 'create'; however, the writer's or translator's internalisation of a metaphor will possibly influence the extent to which it is reinforced in the surrounding context. Clearly, it is by no means the case that every possible collocate and contextual co-referent of a metaphorical item is likely to be compatible with a particular mapping, so the extent to which this is indeed the case may well vary. In the context of translation, depending once again on the degree of the translator's own internalisation of the metaphorical mapping, each individual item may well be handled in isolation, some being reduced, others augmented and still others replaced with different items without reference to a particular metaphorical mapping, with the result that other alternative metaphorical mappings may simultaneously be activated. The result of this is likely to be varying degrees of heterogeneity in terms of the text's metaphorical content.

The listings that follow are designed to illustrate the manner in which strings of collocates and close and approximate contextual co-referents of 'BioBricks', or its equivalent term, can vary between original and translation. They include every instance of each item listed, in the order in which they occur within the immediate vicinity of a mention of the term in question, or when BioBricks inventor Drew Endy is describing what BioBricks are and how they work. For these source text/target text pairs have deliberately not been lined up in order to ensure that it is possible to view the data globally and that the emphasis remains on larger metaphorical structures. While the disadvantage of this is that the exact translation of each relevant item is not tracked, on the positive side, the global view of the data that it provides should bring any shifts in metaphorical emphasis into sharp relief.

English: built—piece of DNA—parts—assemble—pieces—parts—building—parts—section of DNA—parts—fabricated—stitched together—larger bits of DNA—components—standardized parts—Standard Biological Parts—parts—part—built (Gibbs 2004:76–8)

Comment: Besides the term itself, the GENETIC MATERIAL IS BRICKS metaphor is activated for readers via three instances of the verb 'build' and, less unequivocally, by 'fabricated'. The phrase 'stitched together'—applied to the BioBricks themselves—is incompatible with this metaphor, as it evokes the GENETIC MATERIAL IS FABRIC mapping—one that is at best only marginally present in any of the translations. On the other hand, through the presence of items such as 'parts', 'assemble' and 'components', the alternative mapping of GENETIC MATERIAL IS COMPONENTS is in fact more strongly discernible throughout the stretches of discourse in question than that of GENETIC MATERIAL IS BRICKS.

German: Module—vorgefertigten—einbauen—Versatzstücke—zusammenfügen—Bausteine—Versatzstücke—Module—eigens konponierten Stück—Ziegel- oder Bausteine—Module—zusammenmontieren—Konstruktion—kombinieren—bauen—biologische Standardkomponenten—Module (Gibbs 2004: 70–2)

[modules—prefabricated—install—components—assemble—building blocks—components—modules—specifically composed piece—bricks or building blocks—modules—fit together—construction—combine—build—biological standard components—modules]

Comment: Besides a couple of other individual building-related items (*Konstruktion*, *bauen*), the German translation includes a mention of bricks (*Ziegelsteine*). If anything, the BRICKS mapping is about equally prominent in this version of the text as the COMPONENTS one.

Russian: детали—элемент—фрагмент ДНК—конструирование—блоки—детали—элементы—конструктор—детали—сегменты ДНК—детали—создавать—соединять—сегменты ДНК—компоненты—стандартные взаимозаменяемые детали (Gibbs 2004:48–50)

[parts—element—DNA fragment—construction—blocks—parts—elements—construction set—parts—DNA segments—parts—create—join—DNA segments—components—standard mutually replaceable parts]

Comment: The Russian translation adds a rich image of its own—that of the children's construction set. Apart from that, there are very few references to building (only *конструирование* and *блоки*), with the COMPONENTS mapping once again predominating.

Polish: zbudowali—fragment DNA—budowanie—budować—części—części—części—fragmenty DNA—elementy—produktowane—łączyć—fragmenty DNA—części—standaryzowane części—Standardowe Części Biologicznych—części (Gibbs 2004:60–2)

[built—DNA fragment—building—build—parts—parts—parts—DNA fragments—elements—produced—connect—DNA fragments—parts—standardised parts—Standard Biological parts—parts]

Comment: Here there are three instances of building (zbudowali—budowanie—budować), but many lexical items that conform to the COMPONENTS mapping, once again.

Italian: comporre—scambiare—costruiti—costituito—un tratto di DNA—componenti—assemblaggio—parti—componenti—costruzione—parti—sezione di DNA—componenti—fabbricati—uniti—frammenti più grandi di DNA—componenti—componenti standardizzati—fabbricare—componenti biologici standard (Gibbs 2004:76, 78, 80)

[put together—exchange—constructed—constituted—a stretch of DNA—components—assembly—parts—components—construction—parts—DNA section—components—built—joined—larger fragments of DNA—components—standardised components—build—standard biological components]

Comment: The Italian perhaps contains slightly less emphasis on building (costruiti—costruzione and, to a lesser extent, fabbricati—fabbricare), while the COMPONENTS mapping comes very much to the fore.

Chinese: 製造—段 . . . DNA—組件—做出—組件—拼裝-做出—建構—組件—製造—接合—組裝—組件—組件一標準化的基因組件—製造-組件—標準生物組件-組件—拼裝—做出 (Gibbs 2004:90–2)

[manufacture—section . . . of DNA—component—issue—component—assemble—issue—construct—component—manufacture—connect—assemble—component—component—standardised genetic components—manufacture—component—Standardised Biological Components—component—assemble—issue]

Comment: The sense of building is only present in the word 建構 ('construct'), which occurs only once. Besides this, it is the COMPONENTS mapping that completely predominates.

As suggested earlier, if we look at individual items from this collection of words, we are likely to find some specific translation shifts (such as the rendering of *assemble* as *конструирование* ('construction') in Russian). When focusing on the collection in its entirety in each language, on the other hand, we observe the more global reconfigurations that are noted in the previous comment sections. Drawing the six sets of language-specific observations together, what can be observed most clearly is a) a varying of the relative activation of the BRICKS and COMPONENTS mappings from German (where BRICKS is numerically most significant) to Chinese (where it is least significant); b) the addition of the 'construction set' rich image in Russian, which indicates that an existing image can be developed in translation; and c) the removal of the FABRIC mapping (as instantiated by the English item 'stitching') in all target languages. In other words, the extent to which the language of building is activated varies from one target language to the next.

Overall, none of the versions support the term's rich image in a fully homogenous manner—and this includes the source text. Some items indeed relate specifically to the domain of bricks and building that is evoked by this term, but others inevitably do not (or else relate to a different domain). As the listings show, other items will represent alternative mappings, which may be relatively more compatible (e.g. BRICKS vs. COMPONENTS) or more incompatible (e.g. BRICKS vs. FABRIC). This situation gives rise to what may with full theoretical force be termed a mixed metaphor, even if in some cases the mix is more extreme than in others. As already noted, the more extreme of the mixes (i.e. BRICKS vs. FABRIC) is present only in the source text, although it is not possible to conclude whether the universal removal that it undergoes in translation is a typical way for translators to deal with this phenomenon. Furthermore, given that it has not been retained in any version, can we conclude that the translators have achieved an understanding of the underlying metaphorical dynamics of these passages of text? I believe the answer to this latter question must be a qualified 'yes'.

Of all the factors that contribute to the cognitive impact made by a metaphorical term like this, the manner in which the mapping is reflected in collocations and contextual co-referents is not to be ignored. The multilingual data of the extended example that has been discussed reveals a degree of variation in the level of metaphorical heterogeneity within relevant stretches of text. However, the issue of contextual heterogeneity is certainly not specific to metaphorical terminology, as one can no doubt expect it to be of relevance to the communicative force of all types of metaphorical expression. In this particular instance, there is some evidence of the mapping having been broadly internalised as translators bring less metaphorically compatible items into line with the mapping that underlies the item of terminology—an approach that implies a certain degree of intentionality given the universality of its adoption.

This detailed analysis of the translators' treatment of the textual material that surrounds this theory-constitutive term, when combined with the questions of term transparency and the approaches to mediating the term's newness that were discussed earlier, suggests that translators can take considerable trouble to arrive at a terminological form that makes the metaphorical meaning of the original term available—to the extent of becoming the agents of secondary term formation (in the case of both BioBricks and, to a lesser extent, wetware) and sometimes even including two alternative forms in parallel—and indicates that larger metaphorical structures potentially come into play even when an item as seemingly free-standing as a terminological unit is being translated. While the multilingual data reveals considerable variation in the specifics of the translators' approaches, the picture that emerges is of translators who are active term creators and committed mediators of meaning.

That said, the overall care that the translators take is not likely to be so great if a metaphorical term is more informal in nature. Whether or not mixed metaphors contained in the environs of a terminological metaphor (or other type) tend to be removed in translation, it has been my view for some time that further work on the phenomenon of mixed metaphor is needed, in both translation studies and more generally within metaphor research where few studies have so far been attempted (but see Gibbs 2016).

5.1.3.2.5 THEORY-CONSTITUTIVE

The 15 examples of theory-constitutive metaphorical expressions that I was able to identify typically refer to genetic material as components along the lines discussed in the previous section and to DNA as code. Along with terminological and exegetical expressions, it is the category whose members are best preserved in translation—to such an extent, in fact, that hardly a single expression is shifted into another category. The small number of examples available means that the main repeated translation procedure that can be observed is that of a fairly basic retention that sometimes includes rewording with synonyms; there are also a number of instances of explicitation, as one would expect. Both of these are illustrated in the following example:

English:	**standardized parts** (Gibbs 2004:77)
Russian:	**У стандартных взаимозаменяемых деталей** (Gibbs 2004:50) **[Standardised, mutually replaceable components]**
Polish:	**standaryzowane części** (Gibbs 2004:62) **[standardised parts]**
Chinese:	**標準化的基因組件** (Gibbs 2004:90) **[standardised genome components]**

Explicitation can be seen in the Russian and Chinese translations, while in the Polish we see no such changes introduced.

5.1.3.3 Conclusion

This section has been mainly devoted to a discussion of how the translation of two items of metaphorical terminology raises issues of a textural nature and shows the extent to which translators consciously act as mediators even when rendering items of terminology, which would normally be considered too 'pre-packaged' to permit this to happen. Besides that, with this parameter, there is a significant tie-in with the scientific content of the articles and what we seem to find is that there is generally less room for rewording when metaphorical expressions that are felt to be central to this content are translated.

5.2 Conventionality

I have devoted considerable space to my discussion of metaphorical purpose, and this new section is going to be shorter. After an overview of some of the procedures that are used in the context of this new parameter, I will consider the light that the data from this part of my investigation (and elsewhere) casts on Kloepfer's boldness hypothesis.

5.2.1 Background

This broad notion includes such informal concepts as strength, frequency of use, pervasiveness, embeddedness and vitality. It describes the extent to which a metaphorical expression has become a standard, unremarkable part of the 'lexicogrammatical patterns' (Caballero 2003a:104) of a language as well as the degree to which it may have lost—or retained—its original metaphorical force (Croft & Cruse 2004:204).

There are a number of different taxonomies that have been suggested for this area. Goatly distinguishes a total of five levels of metaphoricity—*dead and buried*, *dead*, *sleeping*, *tired* and *active* (1997:32)—with the first four denoting different degrees of lexicalisation. Hanks argues that the smaller the number of semantic features shared by source and target concepts the greater the metaphoricity (2006:31). In addition, he asserts that abstract nouns are 'not used metaphorically at all' (2006:32), although interestingly it seems that the data analysed in this study contains at least one or two examples that seem to contradict this statement (for instance, *a diverse population of glial cells*: Fields 2004:55).

Newmark, on the other hand, famously identifies the following six kinds of metaphor:

> *dead*: 'where one is hardly conscious of the image' (e.g. *space*, *field*, *line*, *top*, *bottom*)
>
> (1988b:106, emphasis original)

cliché: 'metaphors that have perhaps temporarily outlived their usefulness, that are used as a substitute for clear thought' (e.g. *break through*, *a jewel in the crown*)

(1988b:107, emphasis original)

stock: 'an established metaphor which in an informal context is an efficient and concise method of covering a physical and/or mental situation both referentially and pragmatically' (e.g. *I can read him like a book*, *keep a straight bat*)

(1988b:108–9, emphasis original)

adapted: 'a metaphor that has been altered from its standard form' (e.g. *the ball is a little in their court*)

(1988b:111, emphasis original)

recent: 'a metaphorical neologism, often "anonymously" coined, which has spread rapidly in the SL' (e.g. *with it*, *groovy*)

(1988b:111, emphasis original)

original: 'contain the core of an important writer's message' and 'are a source of enrichment for the target language'; this category also includes 'original or bizarre metaphors in "anonymous" non-literary texts'

(1988b:112)

To this, Dickins adds the notion of conceptual metaphor (which he terms 'schematic') as a second dimension that can potentially underlie a metaphorical expression of any of the aforementioned types (2005:247). At the same time, he removes cliché metaphors, because he argues that they are defined 'aesthetically, rather than in terms of metaphorical force' (2005:238) and also places brackets around adapted metaphors, which he considers to be 'a somewhat odd category' (2005:242). What he is left with is a series of five categories arranged in order of their metaphorical forcefulness—*dead, stock, recent, (adapted), original*—all of which can be potentially enhanced by the presence of a schematic dimension (2005:247).

In a study closely based on corpus data, Deignan suggests the following four categories, partly on the basis of Lakoff's and Goatly's work (2005:39):

historical: words such as *comprehend* (based on a Latin word meaning 'to take hold thoroughly') and *pedigree* (an image metaphor based on the French for a crane's foot)
dead: *deep* (of colour), *crane* (the machine)
conventionalised: *whisper* (of wind), *barrier* (e.g. to understanding)
innovative: any metaphorical expression that is new or original

Deignan suggests the following criterion for distinguishing between conventionalised and dead: 'Where a literal sense of a word is perceived as more core than an established metaphorical sense, the second sense is regarded as a conventionalized metaphor', whereas where there is no such apparent relationship of coreness and dependency, the metaphor is considered to be dead (2005:42). In addition, on the basis of corpus evidence, she observes that when used in a conventionalised metaphorical sense, lexical items will generally be qualified by other target domain words (2005:42–3: e.g. *the machinery of government* vs. *expensive new machinery*, and *starved of weapons* vs. *starving*).

Müller recommends decomposing the traditional concept of what she terms *vitality* into the separate notions of *conventionalisation*, *consciousness* and *transparency* (2008:184). However, taking issue with Lakoff and Turner's three-way categorisation of metaphors as either *historical*, *entrenched* or *novel* (1989:128–31) and building on an unpublished manuscript by Kyratzis, Müller proposes a 'single dynamic category' of metaphors (2008:189) on the basis that so-called dead metaphors can easily be activated if a speaker or writer wishes to utilise them in a novel manner, activation being effected by means of a number of textual devices such as adverbial modification and repetition (Kyratzis 2003, quoted in Müller 2008:190–2). What remains, however, is a notion that all but the deadest, most historically embedded metaphors should be considered simply as more or less frozen or fresh (Kyratzis 2003, quoted in Müller 2008:193). Müller's conclusion is to propose a single dynamic category that 'covers a spectrum ranging from sleeping to waking metaphors' based on the degree of activated metaphoricity (2008:195). I will draw back from such a radical solution, as it does not yield categories that provide the contrasts needed for us to try to discover more about metaphor in translation from the starting-point of this important concept.

5.2.2 *How These Concepts Feature in the Data*

While it is clear that some metaphors are stronger than others—and that their presence in a text or translation is of inherently greater interest—defining what relative metaphorical 'strength' consists of in a precise manner is not easy. A single parameter is, however, being used to reflect a range of interrelated notions (such as strength, frequency of use, pervasiveness, embeddedness and degree of lexicalisation) that could together be referred to as 'conventionality'. To rule out possible subjectivity, great efforts have been taken to cross-check initial categorisations against the extent to which different metaphorical usages were attested in Internet texts written in the various languages included in the research.

Because of the compromise they represent between specificity and generality, the values for this parameter are based on those proposed by Deignan in the previous section, but with the addition of a category for 'grammaticalised' expressions. This consequently gives rise to the following five categories arranged in approximately decreasing order of conventionality.

1) Items such as *axon* or *dendrite*, the metaphorical meanings of which are concealed because the item has been formed on the basis of roots that derive from languages no longer spoken, are classified as 'historical'.
2) Words such as *trigger* (e.g. an effect), *spur* (a process), *hooked* (on drugs) or *crack* (a code), the meanings of which exist in parallel with alternative, literal meanings, are understood as examples of 'dead' metaphor.
3) Expressions whose metaphoricity is dependent on the precise grammatical form in which they appear are referred to as 'grammaticalised'. For example, the expression *nature's code* can be considered an instance of humanisation because of the notion of ownership that is entailed (see Interlude One, where this was discussed in detail). In most cases, the metaphoricity of such expressions is fairly marginal. This concept should not be confused with Halliday's notion of 'grammatical metaphor', which is something totally different (see Halliday 1985:319–45 and Romero & Soria 2005).
4) In practical terms, for an expression to be classified as 'conventionalised' at least one other instance of a similar usage will generally need to be located elsewhere. In many cases, evidence of such a usage on the Internet has been detected through use of the Google search engine. In practice, this category is the most frequent and accounts for the majority of examples of each of the five purpose categories, for example.
5) An expression can be considered 'innovative' for a number of reasons: for example, it is an original expression within a particular language (*tiny cellular butlers*—Freeland & Hurst 2004:86); it reflects original scientific thought (*Life, version 2.0*—modelling life on a computer program: Gibbs 2004:76); it is an original metaphor relating to a conventional mapping (*glial cells merely nurtured them*—which derives from the mapping BRAIN CELLS ARE SENTIENT: Fields 2004:56); or it is a standard expression used (appropriately or inappropriately) in an original context (the Polish *często mieszając słowa jak groch z kapustą* = 'frequently jumbling words in a disorderly manner' or, literally, 'like peas with cabbage': Javitt & Coyle 2004:37). In general terms, the richer the image, the more likely it is to be innovative. Many extended metaphors also tend to be innovative.

Together, these five categories are able to account for every example contained in my data.

Finally, it should be pointed out that the notion of 'boldness' does not form the basis for a separate parameter. Rather, this complex concept—which on the face of it would seem to entail a combination of vividness and originality—appears to be chiefly reflected in the richness (see Section 6.1) and conventionality parameters. Interestingly, however, the presumption cannot be made that most bold metaphorical expressions will be both rich and innovative. There is in fact no major conjunction of parameters here, as they seem to conflict with each other in what would appear to be a counter-intuitive manner: not all rich images are innovative, neither are all innovative expressions rich images. So boldness is not unequivocally reflected in the parameters that have been selected for the study.

5.2.3 Translation Procedures and Analysis of Examples

As we will see next, there is perhaps slightly more crossover between these categories than is the case with most or all other parameters. As with most of the subsections of Section 5.1.3.2, this section is shorter because it also anticipates the more detailed treatments offered in Chapter Six.

5.2.3.1 Numerical Background

As usual, this section focuses on a comparison between how expressions are distributed across the various categories in both source and target texts, which is as set out in Table 5.2:

Table 5.2 A Comparison of the Numbers of Metaphorical Expressions in Each Conventionality Category in the Source and Target Texts

Source language conventionality categories (numbers of source language expressions shown in brackets)	*Target-language conventionality categories (all six target languages taken together)*				
	innovative	*conventionalised*	*dead*	*grammaticalised*	*historical*
innovative (36)	121	16	0	0	0
conventionalised (212)	36	688	2	0	1
dead (8)	0	3	5	0	1
grammaticalised (10)	1	13	0	11	0
historical (2)	0	3	0	0	6

It is interesting to note the vast range of frequencies, with conventionalised expressions accounting for around 80% of the total number of examples and occurring approximately six times more frequently than innovative expressions, the next most common category. At the other end of the spectrum, there are only two examples of historical metaphorical expressions.

The translation procedures that we see emerging from this table can be summarised as follows:

i) There is a shift from innovative to conventionalised.
ii) There is a converse shift from conventionalised to innovative, which is proportionately much less significant than point i).
iii) Although some dead expressions are revived in translation, members of the category remain largely defunct.
iv) While most grammaticalised expressions remain as such, a reasonable number become conventionalised.
v) The tiny number of historical expressions remain largely as such, but also give rise to three instances of conventionalisation.

Examples of these procedures will be presented in the next section.

5.2.3.2 Examples of the Different Procedures

i) Two instances of an innovative expression becoming conventionalised in translation can be seen in the following:

English:	**nature's skill as a primordial software designer** (Freeland & Hurst 2004:91)
French:	**le savoir-faire de la nature** (Freeland & Hurst 2004:65) [**nature's know-how**]
Italian:	**il talento progettuale della natura.** (Freeland & Hurst 2004:94) [**nature's talent for planning**]

In this example, an unusual insight gives way to two standard formulations.

ii) Next, there are relatively fewer examples of the converse procedure (even if they are more in number). These examples include a number in which a conventionalised expression is rendered, literally, by an innovative one. This occurs either because the expression does not appear to be widely used in the target language (as in the following example) or because the translator seems to have selected something other than the standard translation.

English:	we wanted to know if glia could **monitor** neural activity (Fields 2004:56)
Polish:	Postanowiliśmy . . . zbadać wobec tego, czy komórki glejowe mogą **monitorować** pobudzenia . . . (Fields 2004:29) [We decided . . . to research whether glial cells could **monitor** excitation . . .]

It should be pointed out that I have only marked such translations as innovative if a detailed search has failed to reveal any possible precedents in the relevant language. For example, it is easy to find abundant evidence for the conventionalised nature of the word *monitor* in the expression *glia could monitor neural activity*; however, detailed searching failed to bring to light any parallel usages in Polish, which has led me to conclude that this literal Polish translation should be marked as innovative.

iii) As the table reveals, there are a small number of examples of a dead metaphor being translated as a dead metaphor:

English: pressing a lever hundreds of times for a **single hit** (Nestler & Malenka 2004:80)
German: wenn sie ihn für **den nächsten Schuss** ein paar hundert Mal drücken muss. (Nestler & Malenka 2004:36)
[if it has to press it several hundred times for **the next shot**]

However, as with the few other examples of this phenomenon, what occurs here is the more or less obligatory replacement of a source-language dead metaphor by its natural target language equivalent, which also happens to be a dead metaphor. As such, I believe that it contributes little to our understanding of how scientific metaphor is translated—in just the same way that Catford's system of shifts is not of anything more than historical interest to the discipline (see Section 3.2.1.1).

As stated earlier, there are also a few examples of dead to conventionalised shifts, such as the following:

English: dampen craving long enough for an addict to **kick** the habit (Nestler & Malenka 2004:85)
Polish: tłumiąc głód na tyle, by **wytrącić** narkomana z nałogu. (Nestler & Malenka 2004:61)
[quelling hunger sufficiently to **wrest** an addict from the habit]

These shifts do perhaps bear witness to a larger degree of translator choice than is the case with dead to dead translation.

iv) The next example illustrates all of the procedures for translating grammaticalised expressions that are revealed in the table:

English: **Nature's version** [of the genetic code] looked less elegant than several of the theorists' hypotheses. (Freeland & Hurst 2004:86)
German: **die Lösung, welche die Natur gefunden hatte**, erschien weniger elegant als einige theoretische Vorschläge. (Freeland & Hurst 2004:87)
[**the solution found by nature** appeared less elegant than a number of theoretical proposals.]

Russian: То, **что придумала природа**, выглядело гораздо менее изобретательным, чем многие гипотезы ученых. (Freeland & Hurst 2004:55)
[**What nature thought up** seemed much less inventive than many scientists' hypotheses.]

Polish: **Wariant, którym posłużyła się przyroda**, okazał się znacznie mniej elegancki niż niejedna z hipotez. (Freeland & Hurst 2004:58)
[**The version that nature used** was significantly less elegant than a number of hypotheses.]

French: **la méthode choisie par la nature** était moins élégante que certaines des hypothèses qui avaient été proposées (Freeland & Hurst 2004:60)
[**the method chosen by nature** was less elegant than certain of the hypotheses that had been proposed]

Italian: **La versione elaborata dalla natura** sembrava meno elegante di molte delle ipotesi formulate dai teorici. (Freeland & Hurst 2004:90)
[**The version elaborated by nature** seemed less elegant than many of the hypotheses formulated by the theorists.]

Chinese: 跟科學家提出的幾種假說比較起來，**自然界的版本**實在不怎麼漂亮。(Freeland & Hurst 2004:78)
[compared to several hypotheses proposed by scientists **nature's version** is really not particularly beautiful.]

As discussed in Interlude One, the expression *nature's version* is only considered metaphorical because of its grammatical form—a form that is not retained in any of the translations. In the German, Russian, French and Italian translations, the metaphoricity is made more explicit, with the resulting expression being classified in each case as conventionalised. The Polish translation, on the other hand, can be considered an instance of innovation; what is different about this example is that it represents a mapping—NATURE IS A USER OF CODE—that is believed to be rare and is certainly not found elsewhere in my data. Finally, the Chinese translation is also an example of grammaticalised metaphor, although ambiguously so as discussed in Interlude One.

In general terms, as can be seen in Table 4.4, the mapping NATURE IS A CREATOR OF CODE is very largely dependent on the entailment of ownership that is conveyed by the possessive construction. The previous example is perhaps unusual in that the metaphoricity is preserved in each of the translations, albeit in a range of different manners.

v) Finally, as stated earlier, most historical expressions remain as such:

English: twiglike receptor, or **dendrite** (Fields 2004:55)

Russian: на коротких ветвящихся отростках (**дендритах**) (Fields 2004:24)
[on short branching outgrowths (**dendrites**)]

Polish: z wypustkami przesyłającymi sygnał do komórki, zwanymi **dendrytami.** (Fields 2004:28)
[with appendages transmitting the signal to the cell, known as **dendrites.**]

Italian: un recettore dall'aspetto simile a un rametto, chiamato **dendrite** (Fields 2004:50)
[a receptor similar in appearance to a branch, known as a **dendrite**]

Quite clearly, though, this result is dependent on the reliance of these target languages on Ancient Greek as a source for scientific terminology—a reliance that is of course shared with English.

On the other hand, there are three conventionalised solutions, two of which are shown here:

English: Each neuron has a long, outstretched branch**—an axon** (Fields 2004:55)

Polish: Od każdego neuronu odchodzi długa wypustka, najczęściej rozgałęziona na końcu, zwana **wypustką osiową, neurytem lub aksonem.** (Fields 2004:28)
[From each neuron there emerges a long projection, usually branching at the end, called an **axial appendage, neurite or axon.**]

Chinese: 每一個神經元有一條往外延伸、長長的分支，稱為「**軸突**」(Fields 2004:36)
[Each neuron has an outward extension, a long branch called **an 'axon'**]

Strictly speaking, the Polish translation is mixed, as it retains a historical expression while adding a conventionalised one of its own. The precise meaning of the Chinese term is 'axle protuberance'.

However, like some of the cases involving dead metaphor that were described in point iii), this Chinese translation along with all of those in the previous example are more or less forced on the translator and are consequently of little interest, either within the multilingual context of this research or in translation studies in general. Another observation is that both of the source expressions are terminological in nature—this being a tendency that is likely to be more or less universal.

5.2.4 Kloepfer's Boldness Hypothesis

As was discussed in Section 3.2.2.1, it was Kloepfer's contention that 'the bolder and more creative the metaphor, the easier it is to repeat it in other languages' (1967:116; translation taken from Snell-Hornby 1995:57). As pointed out at the end of Section 5.2.2, boldness is not directly reflected in any of the parameters that this study focuses on; however, there is a clear link between this notion and the rich image and conventionality parameters, even though this connection is by no means absolute. In this short section, we will be looking at different kinds of evidence that may cast some light on Kloepfer's hypothesis to see if it provides confirmation or serves to contradict it. This investigation may of course have three possible outcomes: the evidence may point towards Kloepfer's hypothesis being true, or it may indicate the precise opposite (i.e. that non-bold types of metaphorical expression are in fact the ones that are easier to translate), or it may suggest that bold expressions are indistinguishable from non-bold ones in terms of their ease of translation. For the purposes of the following analysis examples of the last type are described as 'Kloepfer-neutral'.

a) First of all, let us look briefly at three points that emerge from Table 5.2. The first is that conventionalised and innovative expressions seem to be kept within the same category in translation with approximately equal likelihood. Since I take remaining in the same category as a probable indication that an expression did not present a problem to the translator, I would argue that this factor is therefore 'Kloepfer-neutral'.

b) Next, proportionately slightly more innovative than conventionalised expressions keep their metaphorical force in some shape or form in translation. This is definitely pro-Kloepfer.

c) On the other hand, reading between the lines of the table, there appears to be a slightly greater tendency for conventionalised expressions to vanish in translation than is the case with innovative ones. This is clearly another point in favour of Kloepfer.

d) As I mentioned back in Section 3.3.4, when originally coding my data I marked all instances in which I considered that an expression had undergone some fairly significant kind of modification. Although this aspect of the markup has not featured in the book up to now, it may be useful at this point to get an idea about whether the solutions that translators use for translating conventionalised expressions are more standardised than for innovative ones.

A quick count of numbers of conventionalised expressions that have been modified in some way indicates that for every source-language expression there are on average 0.73 instances of modification. This compares to 0.83 for every innovative expression.

Conversely, every conventionalised expression gives rise to an average of 2.70 instances of retention, the corresponding figure for the innovative category being 2.97.

Since the numbers are so close in the two cases, I would suggest that this is another Kloepfer-neutral factor.

e) The corpus contains nearly 30 expressions that are both rich images and innovative in nature. (A theoretical understanding of rich images is not really needed at this stage, although they will be discussed in detail in Section 6.1.) Of these, I intuitively categorised 19 as bold. The following are examples of the kind of expression that is included in this list:

> We would like to be able to routinely **assemble** systems **from pieces** that are well described and well behaved.
>
> (Gibbs 2004:76)

> The current prototypes are relatively primitive, but the vision is undeniably grand: think of it as **Life, version 2.0.**
>
> (Gibbs 2004:76)

> Forming a cell-free void that would **act like a highway separating burning forests on either side.**
>
> (Fields 2004:58)

In addition, I also intuitively identified as bold eight expressions that were either not innovative, or not rich, or neither. Once again, by way of exemplification, this list includes the following three expressions:

> nature needed no **figurative punctuation**
>
> (Freeland & Hurst 2004:87)

> [natural selection] can only **grope** toward an ideal
>
> (Freeland & Hurst 2004:90)

> the glia suddenly began **to light up like bulbs on a string of Christmas lights**
>
> (Fields 2004:57)

A total of 27 English bold metaphorical expressions—out of 268, so in other words approximately 10% of the total number—therefore form the sample that has been used in this section for the purposes of comparison with the average for the data as a whole.

Still working on the assumption that recourse to modification, removal and omission can all serve as textual evidence that the translator was experiencing difficulty in rendering a particular expression, let us now compare the frequencies of these approaches, as well as that of retention (which contrasts with them), for these 27 expressions with those for the whole set of data. If Kloepfer's hypothesis is to receive confirmation then we will expect the 27 expressions to be retained unchanged more frequently than average, and to be modified, removed and omitted more rarely than average.

The distribution of these 4 approaches across all 268 rows of data for all target languages, and also across the 27 expressions, is as set out in Table 5.3:

Table 5.3 Differing Approaches to Translating All Expressions versus Bold Expressions and Their Implications for Kloepfer's Hypothesis (the figures in Column 1 are adapted from Table 3.2), with the 'Retained' category now split into 'Retained' and 'Modified'

General translation approach adopted, and the average number of instances of each per source expression	*All 268 metaphorical expressions; all target languages together*	*27 bold metaphorical expressions; all target languages together*	*Column with the higher value*	*Implication for the hypothesis*
Retained	714	90		
Number per source-language expression	2.66	3.33	2	pro-Kloepfer
Modified	194	15		
Number per source-language expression	0.72	0.56	1	pro-Kloepfer
Removed	293	17		
Number per source-language expression	1.09	0.63	1	pro-Kloepfer
Omitted	290	33		
Number per source-language expression	1.08	1.22	2	anti-Kloepfer

The important numbers in this table are the two figures for each approach per source-language expression. These tell us, for example, that given that 714 examples of retention are found in the entire set of data, this approach is used across the 6 target languages an average of 2.66 times per source-language metaphorical expression. Any values for the number of modified, omitted and removed expressions per source language expression that were lower in the second column than in the first would serve as confirmation for the hypothesis, as would one for retained expressions that was lower in the first column than in the second. As can be seen from the last two columns, three of the four measures provided in this table are in favour of Kloepfer's hypothesis and one is against. We can therefore say that the information in the table is supportive of the hypothesis, although it has to be conceded that the number of bold expressions included in the experiment is relatively low.

So if we put all these factors together we see that points a) and d) are neutral, b), c) and e) are in favour and none are against Kloepfer's boldness hypothesis. So, interestingly, and against my expectation, I have to say, we should conclude that based on my data, the measures I have used are supportive of the hypothesis that bold metaphors are easier to translate than conventional ones. There is of course the proviso that this relates to scientific metaphor only, so the result may be different in the context of an investigation based on literary metaphor, for example.

5.2.5 Conclusion

All in all, it seems that the proportionality between the different categories that I reported on in Section 5.2.3 is broadly preserved in translation, albeit with a certain amount of crossover between the innovative and the conventionalised. We also saw some of the shifts that typically occur when translators are faced with different kinds of expression as viewed from the angle of conventionality.

Following this, I briefly investigated Kloepfer's boldness hypothesis and, based on the assumption that difficulty is reflected in approaches to translation that do not involve retention, I concluded that the available evidence seems to support his central contention.

Interlude Two
Metaphors of Genetics

The data contains far more metaphorical expressions (a total of 68 deriving from 10 mappings) that relate to this subject domain than ones that describe nature (18 distributed over 6 mappings). As with metaphors of nature, it is 'Evolution Encoded' that is richly endowed with such expressions (a total of 34), closely followed this time by 'Synthetic Life' with 29. So in this instance too, the discourse of these articles can be expected to be heterogeneous in nature.

The concept of genetic material gives rise to a group of very common mappings, five of which are present in the top-ten list presented in Section 4.2.3.1. In all, genetic material is mapped in the following ways, arranged in decreasing order of frequency of examples in the English texts:

GENETIC MATERIAL IS TEXT (16 instances: e.g. *a* ***six-letter*** *genetic* ***alphabet***: Gibbs 2004:80)

GENETIC MATERIAL IS MACHINERY (12 instances: e.g. *evolution will tend to break genetic* ***machines***: Gibbs 2004:80)

GENETIC MATERIAL IS CODE (11 instances: e.g. *the rules of* ***life's underlying code***: Freeland & Hurst 2004:86)

GENETIC MATERIAL IS LANGUAGE (10 instances: e.g. *the genetic* ***sentence*** *. . . must first be* ***translated*** *from one system of symbols into an entirely different kind of system*: Freeland & Hurst 2004:86)

GENETIC MATERIAL IS SOFTWARE (9 instances: e.g. *The aim . . . was to* ***reprogram*** *cells*: Gibbs 2004:78)

GENETIC MATERIAL IS CIRCUITRY (3 instances: e.g. ***rewire*** *cells to glow in the presence of a particular toxin*: Gibbs 2004:75)

GENETIC MATERIAL IS INSTRUCTIONS (3 instances: e.g. *how [genes']* ***instructions*** *affect the behavior of the protein molecules they describe*: Freeland & Hurst 2004:84)

GENES ARE HUMANS (2 instances: e.g. *Were [the genes]* ***telling*** *the glia to act*: Fields 2004:59)

GENES ARE ELECTRICAL APPLIANCES (1 instance: *the influx caused . . . various genes* ***to switch on***: Fields 2004:59)

GENETIC MATERIAL IS DATA (1 instance: *This kind of* ***data compression***: Freeland & Hurst 2004:86)

This is very much in line with the findings of Hellsten and Nerlich, who identify the three main source domains of metaphorical expressions used in English-language newspaper articles on synthetic biology that appeared between 2008 and 2010 as books, computers and building/engineering (2011:384). The fact that all but two of the 65 expressions that relate to these concretising mappings occur in the two articles 'Evolution Encoded' and 'Synthetic Life' demonstrates how it is possible for several different models of a particular concept to be used in parallel within individual texts. The purposes for which metaphorical expressions within these mappings are employed are predominantly exegesis (46) and theory constitution (15), with very few examples that are evaluative, terminological or discoursal in nature. This implies that expressions that derive from these mappings tend to be central to the core tasks of communicating science and structuring our understanding of scientific concepts. Interestingly, all the examples of theory-constitutive metaphorical expressions within my entire set of data can be found in this cluster of mappings.

All in all, and possibly because of the purpose behind most of the metaphorical expressions, this appears to be a remarkably stable area in terms of retention levels of both individual expressions and the mappings that they represent. That is not to say that particular mappings are not reduced in specific target languages, as this is something that happens with GENETIC MATERIAL IS LANGUAGE in Russian, for example, through both removal and the shifting of individual examples to other mappings. Indeed, this case provides further evidence of the diversity of the modifications that occur:

English: the cellular machinery **translating DNA's message** (Freeland & Hurst 2004:86)
Russian: клеточного аппарата **декодирования** (Freeland & Hurst 2004:55)
[of the cellular **decoding** apparatus]

Even though *apparat translyatsii* ('transmission/translation apparatus') would have been an acceptable alternative to *apparat dekodirovaniya*, the translator's selection of the latter variant transforms GENETIC MATERIAL IS LANGUAGE into GENETIC MATERIAL IS CODE, and in fact, this is one of two examples where Russian adds an instance of this mapping. In the following example, on the other hand, it is the GENETIC MATERIAL IS LANGUAGE mapping itself that is strengthened:

English: the genetic sentence . . . must first be **translated** from one system of symbols into an entirely different kind of system (Freeland & Hurst 2004:86)
Russian: то генетическое послание можно прочесть, только **переведя его с языка**, использующего одну систему символов, **на совершенно другой язык** (Freeland & Hurst 2004:55)

> [the genetic message can be read only **if it is translated from a language** that uses one system of symbols **into an entirely different language**]

This intensification is caused by a certain degree of explicitation, although what is interesting to observe is the lack of consistency that can characterise a single translator's decisions in this respect, both here and elsewhere.

Only one new mapping appears in the course of translation, and that is GENETIC MATERIAL IS INHERITANCE, four examples of which are added in German, each of which involves the use of a word starting with the morpheme *Erb-* 'heredity':

English: Twisted ladder of DNA (Gibbs 2004:80)
German: »Strickleiter« des **Erbmoleküls** DNA (Gibbs 2004:74)
['rope ladder' of the heredity molecule DNA]

The word *Erbmolekül* is not needed here, as *der DNA* [of DNA] would also have been an acceptable translation. Metaphorically speaking, this is the most obvious innovation identified in any of the target texts. Besides this, nearly all the mappings are reduced through removal and omission in what appears to be a non-consistent manner, as we have seen elsewhere.

6 Lakoff and Johnson's Metaphor Types

As we established in Chapter Three, metaphors vary in terms of the source from which they ultimately derive, which may be our bodies' interaction of the world, our knowledge of how it works or our ability to recognise physical similarity between different objects. Of these different types of metaphor, the first two—based, respectively, on image schemas and propositional knowledge—are both kinds of conceptual metaphor, while the third type—so-called image metaphors—are not. As we saw in Section 3.1.3, the article by Lakoff in which these metaphorical distinctions were first proposed (1987c:194–5) also describes a fourth type, 'Aristotle's metaphor', although this is not being considered here, as it is somewhat artificial in nature, and I was unable to find any examples in the texts that I studied. Taken together, all the above metaphor types are designated as relating to an expression's 'provenance'.

Additionally, metaphorical expressions deriving from all of these sources may be highly detailed and rich in associations or else relatively sparse in both these respects. In this chapter, we will be considering such phenomena and attempting to discover how translators tend to handle examples of each type. The topic of the first section will be the first two kinds of metaphor listed in the previous paragraph, with the third type being investigated in the second section, while the question of richness will also be focused on in the course of Section One.

6.1 Image Schemas and Image Richness[1]

Arguably, the image-schematic layer reveals itself most frequently in simple metaphorical expressions of a prepositional nature (e.g. '*in* time', '*on* the way'), where a particular interaction between physical objects is superimposed onto a relationship involving more abstract concepts. While expressions of this type were certainly present in the texts I analysed, they are not by and large being considered here, as they do not generally require a decision on the part of the translator and so, in my view at least, do not present much interest to translation studies.

6.1.1 *Background*

It is a major contention of at least some proponents of the cognitive theory of metaphor that much of our ability to think figuratively—which after all involves understanding one idea or area of life in terms that are suggested by another—is determined by a physical awareness that was laid down in our minds at a time when we were still very young. While in some cases this insight can lead to a downplaying of the extent to which our metaphorical apparatus can vary along the lines of language and culture, there are many scholars who allow for this kind of variation in the models that they have constructed (as has been seen in Chapter Three). This debate will be looked at in what follows, as it is clearly one that is of vital importance for people who have an interest in translation.

6.1.1.1 *Image Schemas*

The idea of the image schema was first proposed by Johnson (1987:23ff.) and Lakoff (1987a:453ff.) on the basis of empirical work that had been conducted by Talmy (1972, 1975, 1978 and 1983) and Langacker (1976, 1987) from the mid-1970s (Dodge & Lakoff 2005:57). Johnson and Lakoff define image schemas as basic cognitive structures that arise in the mind as a result of our repeated interaction with the physical contours of the world and the fundamental properties that they display. They are therefore central to the theory of embodied meaning that is so closely associated with the names of these two scholars and according to which the experiences and physical sensations that a person undergoes in early childhood—such as falling over, crawling across the floor to reach a favourite toy or being wrapped up in warm, comforting bedclothes—serve as a powerful influence on his or her developing conceptual apparatus.

Image schemas are described as preverbal, mostly non-conscious, and emergent (Johnson 2007:144), rather than innate. In view of the manner in which they arise within our minds, it is argued that they shape a large portion of our conceptual system. Furthermore, in spite of their preverbal nature, scholars reason that they play a major role in the syntax, semantics and pragmatics of natural language (Johnson 2007:145); they can also serve as source domains or otherwise justify metaphorical mappings. Importantly for our study, image schemas are imagistic rather than propositional in nature (Kövecses 2006:207), which means that only a relatively small number of the expressions that are contained in the corpus will be derived from this kind of highly abstract meaning structure. That said, in practice, all expressions deriving from an image schema are not only inevitably expressed in terms of propositions, but, as pointed out by Gibbs and Colston, their image-schematic motivation is only partial (1995:353), which means that mixed-provenance metaphors are theoretically possible too—if difficult to trace all the way back to their underlying image schema(s). Kövecses goes so far as to claim that 'most apparently nonimage-schematic concepts (such as journey) seem to have an image-schematic basis' (2002:38).

According to the literature, one of the main reasons for their great significance is that they offer a solution to the problem of 'symbol grounding' (Kövecses 2006:208): it is image schemas that provide one of the main starting points for chains of signification, or in other words, the source from which the meaning of many different symbols and linguistic expressions ultimately derives. Along similar lines, Soares da Silva describes these structures as 'one of the most productive sources of conceptual metaphor' (2013:28, translated).

In spite of the theoretical centrality of the notion, no comprehensive list of image schemas seems as yet to have been proposed. Exhaustiveness is not claimed by either of the main published lists of image schemas (Johnson 1987:126, reproduced in Kövecses 2006:207–8, and Evans & Green 2006:190), and in any case, as Dancygier and Sweetser argue, the very act of formulating such a list may in fact be futile (2014:23). These lists differ from each other substantially, although by combining them (and eliminating obvious overlap) we arrive at the following 37 items:

ATTRACTION	BALANCE	BLOCKAGE
BOUNDED SPACE	CENTRE-PERIPHERY	COLLECTION
COMPULSION	CONTACT	CONTAINER
COUNTERFORCE	CYCLE	DIVERSION
ENABLEMENT	FRONT-BACK	FULL-EMPTY
INTERACTION	ITERATION	LEFT-RIGHT
LINK	MASS-COUNT	MATCHING
MERGING	MOMENTUM	NEAR-FAR
OBJECT	PART-WHOLE	PROCESS
REMOVAL	RESISTANCE	RESTRAINT REMOVAL
SCALE	SOURCE-PATH-GOAL	SPLITTING
STRAIGHT	SUPERIMPOSITION	SURFACE
UP-DOWN		

A possible generalisation that one may make is that many of the items on this list refer to how entities interact with each other in the physical world. The items on this list are 'neither context-bound, nor specific, nor conscious, nor effortful'; they contain no 'detail' and have no specific knowledge associated with them (Lakoff 1987a:453). In this way, image schemas, either singly or in combination, can provide the 'skeletal structure' for any number of 'rich images' (Turner 1991:57), although sophisticated concepts will generally derive from underlying image schemas by means of complex 'inheritance relations' via a 'network of intermediate mappings' (Evans & Green 2006:301), and their image-schematic derivation will be partial, as already stated (see Section 2.3.3.1).

By way of illustration, the CONTAINER image schema is said to derive from the fact that our own bodies are containers (of blood, organs and, indeed, our minds) and that we in turn exist within other larger containers (such as rooms, cities or states of awe or love). This schema provides the structure for metaphorical mappings such as STATES ARE CONTAINERS ('He's in very high spirits') or THE VISUAL FIELD IS A CONTAINER ('Now I've got you in my sights'), all of which include the elements 'interior', 'boundary' and 'exterior' (Kövecses 2006:209). Likewise, the SOURCE-PATH-GOAL image schema accounts for mappings such as LIFE IS A JOURNEY and PURPOSES ARE DESTINATIONS (Kövecses 2006:210–11).

Kövecses also provides some examples of idioms that derive their structure from one or more image schemas—for instance, *To spill the beans* (CONTAINER), *To hit the bull's-eye* (CENTRE-PERIPHERY, PART-WHOLE and SOURCE-PATH-GOAL) and *All roads lead to Rome* (SOURCE-PATH-GOAL and LINK) (2006:225–6, 355). The fact that the articles contain very few sophisticated expressions that are image-schematic in origin could be partly a function of the fact that they are scientific metaphors rather than expressions that can be traced back to everyday, embodied human experience, at least via relatively few intermediate steps.

Proponents of the concept of image schemas argue that some experimental confirmation of their cognitive reality is now becoming available. For example, as reported by Rohrer, some functional magnetic resonance imaging (fMRI) and event-related potential (ERP) experiments demonstrate that brain activation patterns caused by literal and metaphorical language stimuli are consistent with the image-schema hypothesis (2005:182–6). However, apart from Al-Harrasi (2001), there is little or no work that has come to my attention that offers an analysis of real metaphorical expressions that are largely image-schematic in nature. Gries and Stefanowitsch 2006, for example, do not provide any studies that focus on image schemas to any significant extent. On the other hand, Goatly's work, which is based on the Metalude database, 'reveals that a considerable number of metaphorical themes lack a bodily experiential correlation as their basis, suggesting non-universality': Goatly 2007:7). It is consequently a subsidiary aim of this chapter to attempt an image-schematic analysis of the authentic examples contained in the corpus and to see what, if any, confirmation of the cognitive significance of this concept is provided by an observation of translator practice. In this latter respect, I will aim to build on the work carried out by Al-Harrasi (2001).

6.1.1.2 *Propositional Knowledge*

As stated earlier, image schemas do not lie behind all conceptual metaphors, as is at least tacitly recognised by Johnson (1987:125–6), even if they are situated near the centre of the approach developed by Lakoff, Johnson and a number of other scholars closely associated with them. Although it is impossible to say whether this is typical, in my scientific data at least, those metaphorical expressions that derive from an image schema in fact form a tiny minority of the total.

As stated by Kövecses, a human being's knowledge about the world comes in two forms: image-schematic and propositional, for which reason our metaphors are also largely based on these two types of knowledge (2006:128). An example of the former would be the STATES ARE CONTAINERS mapping mentioned earlier and of the latter THE MIND IS A COMPUTER, derived from our understanding of how computers work (2006). A proposition can be described as 'what is asserted by a sentence which says that something is the case—e.g., that the earth is flat, that bachelors are unmarried men, that two plus two is four, and so on' (Pritchard 2010:3; see also Duranti 1997:28). In science and medicine, propositional knowledge is defined as 'what is publicly available in printed or electronic media such as textbooks and journals, and is communicated in medical education, continuing education, conferences, etc.' (Sadegh-Zadeh 2012:15). In the theory of knowledge, this kind of knowledge is contrasted with knowledge by acquaintance, perceptual knowledge, moral knowledge and procedural knowledge (or know-how) (Sadegh-Zadeh 2012:14–15). In the present context, image-schematic knowledge should also be included in this list, while for the purposes of this research, cultural knowledge may be thought of as a sub-type of propositional knowledge.

6.1.1.3 Richness

This concept concerns whether or not a particular expression can be considered a 'rich image'. All metaphorical expressions involve an 'image' that is more or less rich in details and associations and that determines the expression's appropriateness in a particular context. This becomes clear if we examine a simple ill-formed example in which we can observe that the imagery evoked by different one-word metaphors can potentially be at odds with the topic to which they are applied. The sentence 'The text was swarming with typos' is unacceptable because the phrase *swarming with* evokes an image of groups of animals (and in fact principally insects), people or even inanimate entities involved in some kind of busy, generally directionless movement, which means that its use would be appropriate with groups of *wasps, bandits* or *smoke particles*, for example, but not with *typos*, as they are not busy and do not move. This mismatch of imagery in fact reveals the sentence as the learner English that it is; more acceptable alternatives would be 'The text was riddled with typos' or 'The text was peppered with typos'—the images of which are not contradicted by the inherently static nature of typographical errors.

According to Lakoff, rich images arise unconsciously, in that they do not need to be deliberately called to mind or created and can be stored in the memory for extended periods without effort (1987a:447). They are 'fully detailed' evocations of real-world phenomena that 'have specific knowledge associated with them' (1987a:453). Like image schemas, they are 'neither context-bound, nor specific, nor conscious, nor effortful' (1987a). In contrast to image schemas, their structure is sophisticated and nuanced rather than skeletal, although, as pointed out by Kimmel, the notion of 'richness' should be taken as a relative one, as image schemas and rich images 'flow into one another varying with the amount of detail' that they contain (2002:37). This becomes quickly apparent when it comes to analysing real data.

In spite of the obvious centrality of the notion of the 'rich image', it has remained largely unexploited as a theoretical concept within the cognitive approach to metaphor and has not been developed in a methodical or consistent manner. Indeed, some of the main sources of information even seem to contradict each other. Added to this, there seems to be a certain degree of terminological inconsistency surrounding the concept, while the question of how rich images relate to metaphors based on propositional knowledge rather than on image schemas (see Lakoff 1987c:194–5) has not been explicitly addressed in any of the major sources of which I am aware.

Lakoff (1987a) is one of few major scholars who discusses the topic to any significant extent, and certainly in more detail than Johnson (1987), whose discussion is frequently cited as one of the main authoritative treatments of the concept, not least since it is Johnson who claims to have coined the term (1987:24). However, interestingly enough, nowhere in either source is the term *image* precisely defined or delimited. Lakoff, for example, variously refers to what is presumably the same concept as *image* (1987a:444), *conventional image* (1987a:446) and *conventional rich image* (1987a:446) in the space of a few pages. Turner adds the important clarification that images are experienced in various modalities: 'a visual image of a road, an auditory image of a scream, a kinesthetic image of a pinch, an olfactory image of the smell of pine, and so on' (1991:57). Kimmel makes the common-sense observation that '[most] frequently rich images are of visual nature, but auditory, tactile, kinesthetic, olfactory, or gustatory images are by no means excluded' (2002:36–7).

In addition, the notion is generally associated with the basic level, not least because it is here that a speaker's world knowledge is generally mapped out most clearly in imagistic terms. 'Basic level' is a term used to refer to a conceptual level that occupies a midpoint between the specific and the general (see, for example, Rosch 1978; Lakoff & Johnson 1999:27–8); for instance, items such as *book*, *dog* or *mushroom* exist at the basic level, as opposed, on the one hand, to subordinate concepts such as *primer*, *spaniel* or *chanterelle* and, on the other, to superordinate categories such as *publication*, *mammal* or *fungus*. What matters for us that the basic level is the most important for the purposes of informal categorisation.

Writing about figurative expressions in general, Dobrovol'skij and Piirainen (who also avoid a formal definition or even a detailed description of the concept) argue that an expression's real behaviour is governed by 'more concrete, culture-specific knowledge' rather than by knowledge of a more general kind. In other words, while the same high-level mapping may account for specific metaphorical expressions in different languages, these will still differ from each other in terms of their precise linguistic behaviour (2005:130).

From this it becomes immediately apparent why the notion is potentially of such great significance for studies on metaphor in translation, as it is in the interaction between how a metaphorical expression 'behaves' in the target language and its dependence on another expression existing in the source text that much of the interest for the researcher lies.

The association of rich images with the basic level is by no means exclusive. Some examples are clearly too specific to be associated with this level, while others fit very well:

> Being a member of a culture requires one to have a large stock of such conventional rich images. Americans, for example, tend to have images of Marilyn Monroe and Richard Nixon and Cadillac limousines and the Statue of Liberty. They also have images of horses and cats and roses and bicycles and engagement rings and baseball bats.
>
> (Lakoff 1987a:446)

What these examples have in common, of course, is that they are all highly familiar to most Americans, who will typically possess considerable quantities of largely culture-specific, real-world knowledge about each of these items. Interestingly, all of the examples cited by Lakoff here are based on propositional knowledge rather than image schemas.

However, what is less clear, perhaps, is the degree of overlap that exists between the concept of *richness* and other similar notions such as *specificity* and *strength*. While it seems that a metaphorical expression that contains a very rich image does not necessarily also have to be thought of as a *strong* metaphor, it seems intuitively to be the case that the two concepts are in some way linked—for example, by 'richness' being one of the components that make up 'strength'. This question, which does not appear to have been addressed in the literature, has some degree of functional significance in the present research; it has already been briefly probed in Section 5.2.2.

The final point that would benefit from further specification is the question of how rich images relate to metaphors based on propositional knowledge rather than on image schemas. Even though Lakoff, for example, fully acknowledges the fact that many metaphors derive from propositional knowledge (see Lakoff 1987c:194–5 as well as the tacit recognition cited earlier), throughout most of the literature, rich images tend only to be considered in the context of image schemas. However, the fact remains that the notion of richness with all the characteristics described earlier would appear to be equally valid whether the metaphorical expression derives from an image schema or a series of propositions; what is of course missing in the latter case is the binary contrast that exists in the case of the former (in the sense that image-schematic expressions are either augmented with a rich-image element or they are not, while it is difficult to visualise a metaphor based on propositional knowledge that was totally lacking this dimension). Mercifully, however, this would not appear to be of significance for our purposes, and it is certainly not one of my aims to enter into this particular debate. I consequently adopt the view that, since there is no practical reason why the notion of richness should not apply to metaphors based on propositional knowledge as well as to those predicated on image schemas, all expressions contained in the data can be analysed in terms of this parameter.

The concept of the rich image is one that has not been systematically developed anywhere in the literature, but it nonetheless represents another potentially useful concept, particularly if it is extended to metaphors based on propositional knowledge. Rich images are included in the study for a number of reasons. Firstly, if Dobrovol'skij and Piirainen (2005:130) are correct, they provide a suitable standpoint from which to account for phenomena connected to metaphor in translation. Secondly, they also form a major part of Al-Harrasi's analysis, on which it seems worthwhile to build. Thirdly, from the point of view of metaphor research, it seems that studying whether the richness of an image acts as a significant factor in translators' decision processes would be one suitable way of probing the validity and status of the concept.

6.1.2 *How These Concepts Feature in the Data*

6.1.2.1 *Image Schemas*

While this concept is a largely theoretical construct used in metaphor theory, it will be interesting to see if there is any correlation between a metaphorical expression's derivation from image-schematic knowledge and the translation procedures that are opted for. On the basis of the conceptual metaphor theory, it is possible to hypothesise that because image schemas are claimed to possess a certain universality, we may reasonably expect that the translation of metaphorical expressions that are substantially derived from them will present the translator with fewer problems than metaphorical expressions of a different provenance, and therefore that image-schematic metaphors will be more universally translated because they are at least partly based on non-detailed culturally neutral 'skeletal' mappings rather than ones that have been mediated through language or culture. (The same possibly also applies to some image metaphors, although as we will see later in this section a limited number of the metaphorical expressions with this provenance are in fact highly dependent on culturally derived concepts.)

That said, as observed by Gibbs and Colston (1995:353—see Section 6.1.1.1), it should be remembered that many metaphorical expressions display something of a mixed provenance, and the original image schema can be difficult to discern; in other words, an image-schematic metaphorical expression can only be understood through the contribution of propositional knowledge. This situation is somewhat difficult to take account of in classification terms: what, ultimately, is the difference between an expression based on propositional knowledge and one of mixed provenance in which the image-schematic component is buried under multiple layers of propositional knowledge? Consequently, an expression is only classified as image-schematic if it has a major discernible image-schematic element related to an item on the amalgamated list presented in Section 6.1.1.1. As it turns out, though, most of the items from this list are not represented in my data, which only contains examples of five: SOURCE-PATH-GOAL, LINK, BOUNDED SPACE, CONTAINER and UP-DOWN. Each of these is represented by a small number of examples, some of which are shared by two image schemas.

It could well be that more metaphorical expressions are at least partially derived from an image-schematic element that has been overlaid with rich images and layers of propositional knowledge-based meaning, which substantially obscure the expression's image-schematic origin. Whether or not this is the case, the mixed provenance that seems to be associated with such expressions can make the analysis difficult and involve a significant component of subjective judgement.

All expressions that appear to contain a significant image-schematic element—each of which must be predicated on some kind of physicality—are marked as image-schematic in the corpus. Even so, the proportion of image-schematic expressions is still only about 3.7% (i.e. 10 out of 268). There are a number of possible explanations for this. Firstly, it could indicate that image-schematic metaphorical expressions are actually very few in number in the language as a whole, or at least in most scientific discourse. Secondly, it is possible that few scientific metaphors have a significant image-schematic component. Thirdly, because of the factors described above, it is possible that non-trivial image-schematic metaphors are simply difficult to identify in complex real data—or indeed, the difficulty may be down to my own inexperience. The final suggestion notwithstanding, it is, however, possible that the theory as it is currently stated needs to be made more detailed, or at least adapted to account more closely for real data.

6.1.2.2 *Propositional Knowledge*

This is the category to which the bulk (88.4%, or a total of 237 out of 268) of my English metaphorical expressions belong and is thus very much the 'default'. As with the image-schematic expressions, it is currently uncertain whether this is due to the fact that the corpus is about science, or whether such a proportion would be typical for a general language corpus too. That said, it should be pointed out that most of the examples of general, non-scientific metaphors that are found in the corpus (e.g. 'Freud's claim that we actively repress *unpalatable* information', Solms 2004:85) also fall within this category.

6.1.2.3 *Richness*

Although the notion of the rich image is quite clearly a fuzzy one, for the sake of simplicity, I am not going to define different degrees of 'richness'. Consequently, a metaphorical expression is simply deemed to involve a rich image if it appears to evoke a sophisticated complex of associations, whether these are judged highly developed or only moderately so. A large number of standard lexical items, even when used metaphorically (as in 'neural *network*', Fields 2004:55; 'the *reward* circuitry of the brain', Nestler & Malenka 2004:78), do not generally evoke a particularly rich image, largely because they are insufficiently specified to allow this to happen in the minds of most native speakers of English. On the other hand, items that appear to give rise in themselves to a readily available, clearly defined set of associations (such as '[natural selection] can only *grope toward* an ideal', Freeland & Hurst 2004:90; 'the brain's chemistry and *architecture*', Nestler & Malenka 2004:80) can reasonably be considered to constitute a significantly rich image. Such associations can sometimes be supplied by the context in which a relatively unspecified item occurs: compare, for example, 'biochemical infrastructure that *decodes* DNA' (Gibbs 2004:78), which is not likely to evoke a particularly strong network of associations in the mind of most readers, and 'the entire living world had to be perpetually engaged in frenetic *decryption*' (Freeland & Hurst 2004:86), where an approximately synonymous item participates in a more fully delineated context because of the presence of the qualifier 'frenetic'.

However, most rich images also unequivocally presuppose on the part of the reader a degree of real-world knowledge of a linguistic, cultural or specialist nature ('*a six-letter genetic alphabet*', Gibbs 2004:80; 'Natural selection is a "*blind designer*"', Freeland & Hurst 2004:90; 'a "*Goldilocks*" genetic circuit', Gibbs 2004:78; or 'the cell's *operating system*', Gibbs 2004:80). Many of the analogies encountered in the corpus (for example '*Like a wave of cheering fans sweeping across a stadium*, the calcium waves spread', Fields 2004:60) also entail a very specific rich image. Needless to say, some images can be 'rich' for some readers but not for others ('This kind of *data compression*', Freeland & Hurst 2004:86), depending on the particular specialist or cultural knowledge that they may or may not possess, which means that many images that are predicated on scientific knowledge are considered to be 'rich' for the purposes of this research but might be categorised differently in the context of more general language use. In addition, some idioms are described by Lakoff as 'imageable' (1987a:447); these give rise to mental images that are not only surprisingly detailed but also remarkably consistent in terms of the precise detailing visualised by individual native speakers (for example, *to keep someone at arm's length* and *to spill the beans*: 1987a:447–51). Obviously, very many (but probably not all) image metaphors involve an interaction between rich images.

Not surprisingly, metaphorical expressions that may be thought of as 'bold' tend to be categorised as rich images. Notions such as boldness and strength were discussed in Section 5.2.2.

6.1.3 Translation Procedures and Analysis of Examples

This section attempts to observe the trends that exist in terms of how metaphorical expressions within each of these categories are translated. It is believed to be one of the first translation studies investigations (after Al-Harrasi 2001) to look at metaphors based on image schemas.

6.1.3.1 Images Schematic and Propositional Knowledge-Based Expressions

6.1.3.1.1 NUMERICAL BACKGROUND

Table 6.1 sets out how the metaphorical expressions in the multilingual corpus are distributed across these three categories:

Table 6.1 Image-Schematic, Propositional Knowledge-Based and Image Metaphors across the Seven Languages

	English	*German*	*Russian*	*Polish*	*French*	*Italian*	*Chinese*	*All target languages together*
Image-Schematic	10	2	7	7	6	7	7	36
% Image-Schematic	3.73	1.57	4.61	3.91	4.35	4.32	4.29	3.91
Propositional Knowledge-Based	237	115	130	155	126	143	141	810
% Propositional Knowledge-Based	88.43	90.55	85.53	86.59	91.30	88.27	86.50	87.95
Image Metaphors	21	10	15	17	6	12	15	75
% Image Metaphors	7.84	7.87	9.87	9.50	4.35	7.41	9.20	8.14

In the target-language columns, the 'Image-Schematic', 'Propositional Knowledge-Based' and 'Image Metaphors' rows taken together represent the total number of metaphorical expressions that survived as such in the translations. What is perhaps most striking is the small number of image-schematic metaphors (although let us bear in mind the provisos mentioned earlier) and the not insignificant number of image metaphors contained in the English data, as these results are by no means in line with the usual emphases of conceptual metaphor theory, in which the former occupy a central position and the latter a peripheral one. In addition, nearly 90% of English metaphorical expressions belong to the propositional knowledge-based category, although in all likelihood this is a consequence of the scientific nature of the discourse studied.

These broad proportions are essentially retained in all the target languages, barring a substantial drop in the number of image-schematic expressions in German and image metaphors in French, although in both these cases, the small size of the sample indicates that we should treat this result with caution.

In addition, if we compare the percentages in the 'English' and 'All target languages together' columns of Table 6.1, we see that in each case they coincide to within less than one percentage point of each other; in other words, setting aside the possibly quirky results noted in the previous paragraph, none of the three categories of metaphor are either significantly reduced or increased in number in translation. In the light of this, it can be argued that the table contains some slight evidence against the hypothesis tentatively proposed in Section 6.1.2.1 to the effect that, if the predictions of the conceptual metaphor theory are correct, we should expect metaphorical expressions based on image schemas to retain their characteristics as image-schematic expressions in translation more readily than expressions of a different provenance. More will be said about this presently.

As before, a second interesting angle from which to approach the data is to compare the presence of the different types of expression in the English articles and in the translations as one way of seeing if any significant translation trends come to light. Table 6.2 sets out the relevant figures as previously, just for all target languages together.

Apart from the usual loss of examples through removal and omission, this table reveals a quite significant level of stability in the provenance categories, greater, for example, than that seen for the purpose categories discussed in Section 5.1.

The sections that follow will look in detail at the procedures that translators appear to use when translating different kinds of metaphorical expression. A comparison of the numbers reported below will reveal that they are not exhaustive: this is because some sampling has been used, and it has not been easily possible to assign small numbers of expressions to obvious procedures.

Table 6.2 A Comparison of the Numbers of Metaphorical Expressions in Each Provenance Category in the Source and Target Texts

Source language provenance categories	*Target-language provenance categories (overall)*		
	Image shema	*Propositional knowledge*	*Image metaphor*
Image Schema (10)	32	4	0
Propositional Knowledge (237)	4	792	0
Image Metaphor (21)	0	0	75

6.1.3.1.2 HOW IMAGE-SCHEMATIC EXPRESSIONS ARE TRANSLATED

Since image schemas are by definition non-detailed skeletal structures, it is not common for a metaphorical expression to be a simple reflection of a single, unadorned image schema. Rather, as discussed in Section 6.1.2.1, one (or more) image schemas may lie behind a particular metaphorical expression, even though it may be expressed in terms of a rich image that can only be completely understood through recourse to propositional knowledge, with the result that it becomes essential to think in terms of a mixed provenance. Indeed, in each of the 10 examples of this type of metaphor that I have found, the underlying image schema is overlaid with propositional and/or rich-image knowledge. For example, for an appropriate reception of 'Oligodendrocyte glia produce myelin that *insulates* axons' (Fields 2004:57)—a metaphorical expression that contains an instantiation of the BOUNDED SPACE image schema—what is needed is not only an innate grasp of the image-schematic implications but also an intellectual understanding of how insulation works. Similarly, the example 'as animals move up the *evolutionary ladder*' (which instantiates the UP-DOWN image schema; Fields 2004:61) relies to a large extent on what the readers know about ladders for it to work.

Many expressions that were image-schematic in nature could have been presented from the articles. However, as stated at the beginning of the chapter, these do not normally form an object of study because they do not represent an arena of translator choice. In this respect, they are similar to the set of obligatory translation shifts proposed by Catford, which were mentioned in Section 3.2.1.1.

With that in mind, none of the examples of image schema–derived expressions presented here are completely clear-cut, although all of them appear to display an image-schematic element that can be characterised as significant. It is no doubt possible to argue the shadowy presence of image schemas in a much greater number of metaphorical expressions. However, in each of the 10 examples that I have identified, there seems to be a clear parallel between the proposition contained in the expression and the fundamental action or physical situation envisaged by one or more image schemas.

In spite of the highly significant place that image-schematic metaphors occupy in the conceptual metaphor theory, there are only 10 English metaphorical expressions in which a major image-schematic element can easily be identified, with these expressions giving rise to a total of 40 target-language examples. By all accounts, this means that the total sample for this particular parameter is not very large.

There are a number of procedures that can be observed in the translation of image-schematic expressions. Most or all of these are specific to these expressions' image-schematic nature. Along the lines of what was proposed by Al-Harrasi (as described in Section 3.2.2.5), a number of these involve the interplay of image schemas and rich images.

i) Firstly, there are 22 examples in which an image-schematic metaphorical expression that is overlaid by a particular rich image (or by specific propositional knowledge) is translated by an expression that involves the same image schema and rich image (or propositional knowledge). As such it broadly equates to Al-Harrasi's procedure 1.4 'Same Image Schema and Rich Image Domains', although his model does not include the concept of propositional knowledge.

The following example appears to be an instantiation of the SOURCE-PATH-GOAL image schema, with the PATH being the *doorways* and the SOURCE and GOAL the *abutting astrocytes*, and also of the LINK image schema as the doorways connect adjacent astrocytes (although we must remember Dancygier and Sweetser's suggestion that it would be wrong to be too dogmatic in thinking in terms of a fixed list of image schemas). Detail is added through the rich image of the doorway and the reader's propositional knowledge of astrocytes. This configuration is retained in two of the target languages:

English: calcium ions or related signaling molecules simply passed through **open doorways connecting abutting astrocytes.** (Fields 2004:57)

Russian: следствием простого перемещения ионов кальция или связанных с ним сигнальных молекул из одного астроцита в соседний **через соединяющие их открытые ворота.** (Fields 2004:27)

[as a consequence of the simple movement of ions of calcium or the signalling molecules that are linked with it from one astrocyte to the neighbouring one **through open doors that connect them.**]

French: les ions calcium ou d'autres molécules de signalisation passaient simplement par **des portes ouvertes connectant les astrocytes contigus.** (Fields 2004:58)
[calcium ions or other signalling molecules simply passed through **open doors connecting adjacent astrocytes.**]

In other words, each of these translations retains not only the image-schematic nature of the metaphorical expression (unsurprisingly: as Table 6.2 shows us, they all do) but also its realisation by means of rich imagistic and/or propositional knowledge-based elements. It is of course difficult to specify the precise demarcation line between rich image and propositional knowledge (and in a sense it does not matter, as what we are interested in is detailing); even if the notion of *astrocytes* is likely to be predicated solely on propositional knowledge (except in the case of the most specialised readers, perhaps), that of *doorways* will probably comprise a mix of rich imagery and propositional knowledge.

ii) Secondly, there are 16 examples of an image-schematic metaphorical expression complemented by a specific rich image (or by specific propositional knowledge) being translated by an expression that involves the same image schema but a different rich image (or propositional knowledge). This corresponds to Al-Harrasi's procedure 1.6 'Using a Different Rich Image That Realises the Same Image Schema Realised by the Rich Image in the Source Text'.

English: calcium ions or related signaling molecules simply passed through **open doorways connecting abutting astrocytes.** (Fields 2004:57)

German: dass Kalzium-Ionen oder Signalmoleküle anderer Art **einfach direkt von Zelle zu Zelle wandern—durch Poren,** worüber benachbarte Gliazellen direkten Kontakt miteinander herstellen. (Fields 2004:51)
[that calcium ions or signal molecules of another kind **simply move straight from cell to cell—through pores** via which neighbouring glial cells establish direct contact with each other.]

Polish: jony wapnia lub podobne cząsteczki sygnalizacyjne przechodzą przez **otwarte, łączące je kanały.** (Fields 2004:30)
[calcium ions or similar signalling molecules pass through **open channels that connect them.**]

Italian: gli ioni calcio, o alcune molecole segnale a essi correlate, erano semplicemente transitati attraverso **passaggi di collegamento tra astrociti confinanti.** (Fields 2004:52)
[the calcium ions, or certain signalling molecules related to these, had simply passed through **connecting passages between adjacent astrocytes.**]

Chinese:	鈣離子或相關信號分子穿越了**毗連星狀細胞的開通管道**。(Fields 2004:38) [calcium ions or correlating signal molecules have passed through **the open pipeline of adjoining astrocytes.**]

In these various translations, the skeletal structure of the SOURCE-PATH-GOAL and LINK image schemas is preserved, although in some languages, it is realised by different rich images: instead of doorways, as in English, we have pores in German, channels in Polish, passages in Italian and a pipeline in Chinese. This is a phenomenon that we also observe in the other examples, although in some of them the modification is to propositional knowledge rather than rich images.

iii) There are two examples of an image schema being replaced by a metaphorical expression based on propositional knowledge. In the following example, it is the SOURCE-PATH-GOAL once again that comes under examination:

English:	Many **Avenues of Attack** (Javitt & Coyle 2004:55)
Russian:	**Массированная атака** на шизофрению (Javitt & Coyle 2004:36) [**Mass attack** on schizophrenia]
Italian:	**Attaccare su più fronti** (Javitt & Coyle 2004:65) [**Attack on more fronts**]

What is noteworthy here is that the translations are similar in meaning but lack an image-schematic dimension, essentially because the notion of 'avenue' is missing.

While there are very few examples of this procedure—just the two detailed in this example, in fact—they at least indicate that there is nothing sacrosanct about image-schematic metaphors; they are treated like any other metaphorical expressions and will not inevitably be retained in all target texts. (A similar conclusion can be drawn from the handful of expressions that are removed and at least some of those that are omitted, for more details of which see the next two points.)

iv) There are 12 instances of an image-schematic expression being removed. This eventuality was described by Al-Harrasi as procedure 4 'Deletion of the Expression of the Metaphor'. No example is being presented here, as no specific points would be illustrated.
v) Along similar lines, there are nine examples of expressions being omitted. This procedure is also envisaged by Al-Harrasi under the same heading. Once again, no example is being presented here.

Interestingly, all ten examples of image-schematic expressions belong to only two existence categories, there being six examples of concretisation and four of reification (although this becomes more understandable when you look at what the other seven categories are). In no case is an image-schematic expression translated by an image metaphor. Finally, none of the metaphorical expressions added in any of the six target languages (14 in all) are classified as image-schematic.

6.1.3.1.3 HOW PROPOSITIONAL KNOWLEDGE-BASED EXPRESSIONS ARE TRANSLATED

Metaphorical expressions in this category involve all figurative configurations other than the basic interactions of physical entities that are entailed by image schemas and the physical resemblances conveyed by image metaphors. For this reason, the category is much larger than the other two, accounting for 88.43% of the total number of English metaphorical expressions in the corpus. Because of this, a sample consisting of 85 English examples (for the most part randomly selected, but including four expressions that contained a significant culture-bound component) was used as the basis for the analysis, along with their translations.

By no means all the procedures identified in the sample should be thought of as being restricted to propositional knowledge-based metaphorical expressions, as most have a more general applicability to all language, whether metaphorical or non-metaphorical. In other words, while they directly concern the nature of metaphorical expressions based specifically on propositional or cultural knowledge, there is nothing particularly exclusive about most of them. In this sense, the possibilities that are outlined in the following paragraphs should be considered the nearest thing that the research has produced to a detailed parameter-neutral list of procedures for metaphor translation. The following procedures are arranged in approximate order of numerical significance. Not all are procedures in the strict sense, but some are rather types of rewording that give rise to particular translation effects.

i) The first is a procedure in which the original knowledge structures are preserved without significant modification. This procedure is the most common—yielding a total of 107 examples across the sample studied—and involves the selection of wording that is at least close enough to the original to ensure that the metaphor and also the mapping are retained. In practice, this can mean using the most obvious translation equivalent for a single-word metaphor—for example, translating *population* by French 'population' or Chinese '人口' (although in practice, the second of these theoretical equivalents does not occur in my data)—or else choosing perhaps an alternative target-language synonym or other related word, as in the following example:

English: glia **sense** axon activity (Fields 2004:57)
German: dass Gliazellen neuronale Impulse **wahrnehmen.** (Fields 2004:50)
[that glial cells **are aware of** neuronal impulses]

Russian: глия способна **распознавать** импульсную активность в аксонах (Fields 2004:27)
[glia are able to **recognise** impulse activity in the axons]

Polish: neuroglej **wykrywa** aktywność aksonu (Fields 2004:28)
[neuroglia **detect** axon activity]

This is considered the default category, partly because it is the commonest, but also because it represents a fairly 'straight' translation from the source to the target language.

ii) In the second procedure, a more neutral wording is opted for, giving rise to a more generalised knowledge structure (6 examples):

English: neurons and glia **engage in a two-way dialogue** (Fields 2004:55)

Polish: neurony i komórki glejowe **komunikują się ze sobą** (Fields 2004:27)
[neurons and glial cells **communicate with each other**]

French: les neurones et les cellules gliales **dialoguent** (Fields 2004:56)
[the neurons and the glial cells **hold a dialogue**]

In each case, a loss of detail can be observed. In addition, however, in many instances (a total of 39), this kind of modification also entails removal of the metaphorical element:

English: glial cells **respond** to **chattering** neurons (Fields 2004:59)

Russian: глиальные клетки способны реагировать на обмен сигналами между нейронами. (Fields 2004:27)
[glial cells are capable of reacting to the exchange of signals between neurons.]

Polish: komórki neurogleju reagują na potencjał czynnościowy neuronów. (Fields 2004:31)
[the neuroglia cells react to the neurons' action potential.]

French: La communication des cellules gliales avec les neurones et entre elles (Fields 2004:60)
[The communication of the glial cells with the neurons and among them]

As can be seen, such instances are much more common than those that do not involve the loss of the metaphor.

iii) The third procedure is paraphrase, which leads to shifts of emphasis in how the knowledge structure is described:

English: one way the brain revises its responses to stimuli as it accumulates experience—**how it learns.** (Fields 2004:58)

German: Da sich beim **Lernen** wahrscheinlich Synapsenstärken verändern (Fields 2004:50)
[Since the strength of synapses probably changes during **learning**]

Russian: один из способов, с помощью которых головной мозг корректирует свои реакции на раздражители по мере накопления опыта—**так происходит процесс обучения.** (Fields 2004:26)
[one of the ways by means of which the brain revises its reactions to stimuli as experience increases—**the learning process happens in this way.**]

Polish: W miarę nabywania doświadczenia następuje modyfikacja połączeń między neuronami, powodując zmienne odpowiedzi na bodźce. Proces ten uważany jest za podstawę **mechanizmu uczenia się.** (Fields 2004:30)
[As experience is gained a modification occurs to the connections between neurons, leading to a change in responses to stimuli. This process is considered to be the basis of **the learning mechanism.**]

In the above examples—and in 9 more similar ones—the metaphoricity is always retained. However, as was the case with procedure ii), paraphrasing a metaphorical expression can often give rise to a non-metaphorical target expression:

English: many came to view **nature's real code** as little more than a random accident of history. (Freeland & Hurst 2004:86)

Russian: Многолетние спекуляции на тему генетического кода закончились, многие сочли, что он—не более чем случайность. (Freeland & Hurst 2004:56)
[Many years of speculation about the genetic code came to an end and many considered that it was nothing more than chance.]

Polish: wielu badaczom prawdziwy kod genetyczny wydał się po prostu dziełem przypadku. (Freeland & Hurst 2004:59)
[to many researchers the real genetic code appeared simply as the result of chance.]

Italian: molti iniziarono a pensare che il codice realmente esistente in natura fosse poco più di un incidente casuale della storia evolutiva. (Freeland & Hurst 2004:91)
[many have started to think that the code that really exists in nature is little more than a casual accident of evolutionary history.]

Each of these renderings reveals significant shifts in the knowledge structure focus, as can be seen in the wording of each of the translation equivalents.

iv) The fourth procedure, with a total of 20 examples, entails changing the mapping by shifting the expression into a different, although congruent, knowledge structure:

English: alternative ways for genes to **spell out** most proteins. (Freeland & Hurst 2004:87)
Russian: гены обладают определенной свободой выбора, **кодируя** данный белок. (Freeland & Hurst 2004:57)
[genes have a certain freedom of choice when **coding** a particular protein.]
French: fournissant ainsi aux gènes plusieurs façons de **définir** la plupart des protéines. (Freeland & Hurst 2004:62)
[thus providing the genes with several ways of **defining** most proteins.]

Here the mapping seen in the English expression GENETIC MATERIAL IS TEXT becomes GENETIC MATERIAL IS CODE in Russian and GENETIC MATERIAL IS LANGUAGE in French.

v) The presence of a metaphor is sometimes signalled, most commonly through the use of inverted commas (8 examples). In other words, the metaphorical expression is explicitly marked as such by some means or other, with the translator underlining in this way the fact that the expression is being meant figuratively and, very possibly, disowning the wording as having originated from him or her. Here the knowledge structure is usually preserved but is more obviously mediated by the translator:

English: But were glia limited only to **eavesdropping** on neuronal activity . . .? (Fields 2004:56)
Russian: Но ограничивается ли участие глии в нервных процессах только «**подслушиванием**» нейронных переговоров? (Fields 2004:24)
[But is the glia's participation in neural processes only limited to "**eavesdropping**" on neural conversations?]
Polish: Czy komórki glejowe ograniczają się jedynie do "**podsłuchiwania**" aktywności neuronów . . .? (Fields 2004:28–9)
[Do glial cells only limit themselves to "**eavesdropping**" on neural activity . . .?]
Italian: Ma le cellule della glia sì limitano a «**origliare**» l'attività neuronale . . . ? (Fields 2004:50)
[But the glial cells limit themselves to "**eavesdropping**" on the neural activity]

This procedure, which seems to be most common in the Chinese data, is generally combined with another one, most usually procedure i). While the commonest way of signalling a metaphorical expression's presence is through the use of inverted commas, the same result can be reached in other ways too (for example, German *quasi* 'quasi'; Russian *своего рода* 'a kind of', etc.). It occasionally happens that an English expression is signalled in the same manner (for example *the newcomers **'captured'** a subset of the tRNAs*, Freeland & Hurst 2004:91; *making the remainder of the 64 possible triplets **'nonsense'***, Freeland & Hurst 2004:87). In such instances, the signalling marker is generally retained in the target languages, although as we shall now see this is not invariably the case.

vi) The sixth procedure is thus the opposite of number v), or, in other words, removal of a signalling marker—an eventuality that is surprisingly common in the corpus:

English:	the newcomers **'captured'** a subset of the tRNAs (Freeland & Hurst 2004:91)
Russian:	Более сложные варианты аминокислот . . . **оккупировали** их тРНК (Freeland & Hurst 2004:61) [The more complex amino acid variants . . . **occupied** their tRNA]
Polish:	[nowe aminokwasy] [z]**dobyły** dla siebie zestaw tRNA (Freeland & Hurst 2004:63) [[the new amino acids] **captured** the composition of tRNA for themselves]
Chinese:	這種新胺基酸會**捕捉**一組tRNA (Freeland & Hurst 2004:83) [these new amino acids could **capture** a set of tRNA]

With 17 examples, this behaviour is in fact slightly commoner than the addition of signalling, which is to my mind a somewhat surprising result. No particular reason for this behaviour on the part of the translator obviously suggests itself, although interestingly it is contrary to my suggestion that translators may try to distance themselves from some metaphorical wordings.

vii) Next, in six examples we can see a shift from a more to a less culture-related form of propositional knowledge:

English:	frequently jumbling words together into a meaningless **word salad** (Javitt & Coyle 2004:50)
German:	eine wirre Reihung von Gedanken und **Sprachfetzen**. (Javitt & Coyle 2004:64) [a confused succession of thoughts and **shreds of language.**]
Italian:	mescolando alla rinfusa le parole in un <u>miscuglio</u> privo di significato (Javitt & Coyle 2004:60) [jumbling the words in a <u>medley</u> deprived of sense]

There are no examples that I have been able to locate of the opposite phenomenon.

viii) This procedure sees the retention of the mapping in spite of significant changes to the wording:

English:	Schwann cells perform myelination **duties** (Fields 2004:57)
Polish:	komórki Schwanna otaczają aksony i **odpowiadają za** ich mielinizację. (Fields 2004:29) [the Schwann cells surround the axons and **are responsible for** their myelination]
French:	où les cellules de Schwann **se chargent de** la myélinisation des neurones. (Fields 2004:59) [where the Schwann cells **attend to** the myelination of the neurons.]

In each of the five examples that I have found, the new wording is due to the decision of the translator as a closer rendition would by no means have been ruled out.

ix) The concept of 'universal deletion' should be mentioned here as well, although it is not a procedure so much as an observed multilingual phenomenon. The purpose of including it is simply to note when a particular metaphorical expression—often with the entire associated knowledge structure—is removed in all target languages (this occurs a total of four times).

English:	Freud's claim that we actively repress **unpalatable** information. (Solms 2004:85)
German:	Freuds Behauptung, wir würden <u>unangenehme</u> Bewusstseinsinhalte aktiv verdrängen (Solms 2004:79) [Freud's claim that we would actively repress <u>unpleasant</u> consciousness contents]
Russian:	другое предположение Фрейда: мы активно вытесняем **неприятную** информацию в область подсознательного. (Solms 2004:58) [Freud's other presupposition: we actively force out <u>unpleasant</u> information into the realm of the subconscious.]
Polish:	że—jak twierdził Freud—świadomie tłumimy <u>niemiłe</u> dla nas informacje. (Solms 2004:77) [that—as Freud maintained—we consciously repress information that is <u>disagreeable</u> for us.]

French: comme Freud l'affirmait, que nous réprimons les informations désagréables. (Solms 2004:79)
[as Freud affirmed, that we repress disagreeable information.]
Chinese: 佛洛伊德的主張：我們會主動壓抑不愉快的訊息。(Solms 2004:50)
[Freud's assertion: we can actively repress disagreeable information.]

In this example, the metaphorical expression is reflected in each target text but is rendered by non-metaphorical means. There are no examples of universal omission.

Finally, a third group of examples are removed or omitted in every target language, although which of the two procedures occurs varies from language to language.

There are relatively few such examples. Had the phenomenon been more widespread, it might have indicated that certain metaphors or metaphorical expressions (or even single lexical items used metaphorically) were in some ways inherently resistant to translation. As it stands, however, there is insufficient data to allow us to make such generalisations.

x) The sample contains four examples of explicitation, including the following:

English: **assemble** systems **from pieces** (Gibbs 2004:76)
German: **aus definierten, wunschgemäß arbeitenden Modulen** biologische Systeme zu **konstruieren** (Gibbs 2004:70)
[to **construct** biological systems **from defined modules that work as required**]

In each example of these, the original expression is spelled out by the insertion of extra words.

xi) Finally, the following is a (non-exhaustive) list of miscellaneous procedures that are also represented in very small numbers and should be mentioned briefly:

- increasing the level of formality (five examples)
- decreasing the level of formality (one example)
- using a synonym instead of the most obvious translation equivalent (four examples)
- intensifying the original expression (two examples)

It is not believed that any of these are particularly significant but are only ever likely to account for small numbers of expressions.

6.1.3.1.4 CONCLUSION

As stated earlier, a large proportion of the procedures and other translation effects listed previously are not unique to the propositional knowledge-based expressions that are contained in the data. However, taken together, they provide a relatively detailed—but probably not exhaustive—picture of the shifts that occur within what is by far the largest provenance category that is reflected in the data.

6.1.3.2 Richness

The corpus contains a total of 108 English rich images, so this is a kind of metaphorical expression that accounts for almost exactly 40% of metaphorical expressions. Of these instances, slightly more than half (i.e. 57 expressions) are categorised as either exegetical or theory-constitutive, which suggests that rich images might be a particular feature of scientific language (or at least of that of popular science).

As expected, this parameter also undergoes modifications in the translation process. Metaphorical expressions may be omitted or removed, of course, or may retain, acquire or lose a degree of richness:

English:	as animals move up the **evolutionary ladder**. (Fields 2004:61)
German:	Je **höher** in der Evolution Tiere stehen (Fields 2004:56)
	[The **higher** animals stand in evolution]

In this example, a sophisticated, culturally and scientifically determined metaphorical expression in which the notion of upward motion is filled out via the image of a ladder is replaced by a blander rendering in which both these elements are removed, with the source-language expression being translated by one that simply indicates relative positioning. This is, of course, just one of several possible complex modifications that may occur in translation. The purpose of this section is to observe the cross-language tendencies that obtain and to examine precisely what kinds of rewording are brought about by a modification within this parameter. Following that, we will return to Al-Harrasi's classification of translation procedures in order to examine whether it can be usefully developed in the light of the scientific metaphors that are being investigated in this study.

6.1.3.2.1 NUMERICAL BACKGROUND

First of all, details of the total number of rich images and non-rich images (a formulation that is intended to imply 'something that is not a rich image' rather than 'an image that is not rich') in each of the seven languages and how these average out when taking all the languages together can be found in Table 6.3:

Table 6.3 Rich Images and Non-rich Images across the Six Languages

	English	*German*	*Russian*	*Polish*	*French*	*Italian*	*Chinese*	*All target languages together*
Rich Image	108	56	50	61	31	57	57	420
% Rich Image	40.30	44.09	32.89	34.08	22.46	35.19	34.76	35.29
Non-rich Image	160	71	102	118	107	105	107	770
% Non-rich Image	59.70	55.91	67.11	65.92	77.54	64.81	65.24	64.71

As was the case in Table 6.1, this table simply reports the numbers of rich images and non-rich images that occur in each of the languages and the average of their occurrence in all the target languages. As before, percentages are provided for each figure. As we can see from the first column, the ratio of rich images to non-rich images is approximately 40:60 in the English data. The proportion of rich images rises slightly in German but falls in all the other target languages; in each case, the fall is greater than the relatively minor increase seen in German, while in French, what can be seen is a significant drop in the percentage of rich images. Overall, however, we can observe a slight swing towards non-rich images in the target languages (35:65 as opposed to 40:60 in English). In other words, in the translated texts, rich images occur with a slightly lower frequency than in the source texts, as is reflected in the figure for all the target languages taken together.

To make more sense of the processes that give rise to the figures in the final column, we need to look at Table 6.4, which, like the other similar tables that have appeared in previous sections, tracks how the two sets of English examples identified as rich images and non-rich images are numerically redistributed in translation:

Table 6.4 A Comparison of the Numbers of Metaphorical Expressions in Each Richness Category in the Source and Target Texts

Source-language richness categories (numbers of source-language expressions shown in brackets)	*Target-language richness categories (overall)*	
	Rich image	*Non-rich image*
Rich Image (108)	293	57
Non-rich Image (160)	15	543

The overwhelming majority of source-text rich images and non-rich images remain within their respective categories (if they are not removed or omitted, which is not reported here), while in numerical terms it appears that non-rich images are slightly more stable in translation than rich images and that non-rich images tend to be replaced by rich images less frequently than the other way round.

6.1.3.2.2 EXAMPLES OF THE DIFFERENT PROCEDURES

In what follows, the four translation tendencies illustrated in Table 6.4 will be considered in turn, starting with the two smaller categories; the comment about exhaustiveness that I made at the end of Section 6.1.3.1.1 also applies here.

i) First, there are 57 examples of a rich image being replaced by a non-rich image. This modification assumes a wide range of different forms:

a) Use of a word that is less immediate to most people's experience (a total of four examples):

English: calcium ions or related signaling molecules simply passed through **open doorways connecting abutting astrocytes.** (Fields 2004:57)

Polish: jony wapnia lub podobne cząsteczki sygnalizacyjne przechodzą przez **otwarte, łączące je kanały.** (Fields 2004:30)
[calcium ions or similar signaling molecules pass through **open channels that connect them.**]

Chinese: 鈣離子或相關信號分子穿越了**毗連星狀細胞的開通管道**。(Fields 2004:38)
[calcium ions or correlating signal molecules have passed through **the open pipeline of adjoining astrocytes.**]

In this example, the translators replace the word *doorways* with *channels* and *pipeline*, which are likely to be further removed from most readers' experience and will therefore enter into fewer networks of mental associations. An *open doorway* is in itself deemed to be a rich image simply because of the everyday familiarity that all readers will have with it and that engenders a number of clear associations.

b) Use of a more abstract word (three examples):

English: Each neuron has **a long, outstretched branch** (Fields 2004:55)

Polish: Od każdego neuronu odchodzi **długa wypustka, najczęściej rozgałęziona na końcu** (Fields 2004:28)
[From each neuron there emerges **a long projection, usually branching at the end**]

Italian: Ogni neurone ha **una lunga ramificazione molto estesa** (Fields 2004:50)
[Each neuron has **a long, very extended ramification**]

Both translators replace the concrete *branch* with abstract alternatives in which the metaphor becomes impoverished; this is consistent with the general weakening of the mapping NEURONS ARE TREES that we saw in the discussion in Section 4.2.3.2.2.

c) Partial deletion of the metaphorical expression. This may occur by partial removal of the expression (five examples):

English: researchers **mapped** the **regions** of the brain (Nestler & Malenka 2004:81)
Russian: исследователям удалось идентифицировать **области** головного мозга (Nestler & Malenka 2004:52)
[researchers managed to identify **regions** of the brain.]

Although the SCIENTISTS ARE CARTOGRAPHERS mapping is essentially lost, there remains an element of the original metaphorical expression.

It can also occur by partial omission (one example):

English: But those who become addicted to cocaine **sprout additional spines on the branches** (Nestler & Malenka 2004:83)
Russian: У животных, пристрастившихся к кокаину, **шипиков на дендритах** гораздо больше (Nestler & Malenka 2004:55)
[Animals addicted to cocaine have far more **spines on their dendrites**]

The translator omits the item *sprout* and converts *branches* to *dendrites*. Once again, this appears to be in line with the weakening of the mapping NEURONS ARE TREES.

d) Shift from rich image to non-rich image in conjunction with a change of mapping (33 examples):

English: **nature's skill as a primordial software designer.** (Freeland & Hurst 2004:91)
(NATURE IS A SOFTWARE DESIGNER)
French: **le savoir-faire de la nature** (Freeland & Hurst 2004:65)
[nature's know-how]
(NATURE IS SKILLED)
Italian: **il talento progettuale della natura.** (Freeland & Hurst 2004:94)
[nature's talent for planning.]
(NATURE IS A PLANNER)

Here the specific knowledge that most readers will associate with software designers is dissipated in each case.

e) Replacement of a vivid word with a more neutral one (11 examples):

English:	if glia could **chatter** (Fields 2004:56)
French:	si les cellules gliales **communiquent** (Fields 2004:57) [if the glial cells **communicate**]

This example typifies this kind of replacement very clearly. (The same thing happens in the German, Polish and Italian translations of the same example.)

f) Generalisation (five examples):

English:	as animals move up the **evolutionary ladder**. (Fields 2004:61)
German:	Je **höher** in der Evolution Tiere stehen (Fields 2004:56) [The **higher** animals stand in evolution]
Chinese:	越是**高度**演化的動物. . . (Fields 2004:43) [the **more highly evolved** an animal. . .]

This example, which was already quoted at the beginning of Section 6.1.3.2, demonstrates how the original metaphorical expression can sometimes be retained only in outline, with a detailed notion being substituted by a more general one.

The common theme across all of the six different procedures just described is quite clearly a loss of detail and/or specificity in the metaphorical expression. As already indicated, shifts in this direction are relatively numerous in the data.

ii) With only 15 examples, movement from non-rich image to rich image is less common by far than the type considered in point i). These shifts seem to be particularly associated with two different specific types of rewording:

a) Use of a more vivid, specific image (six examples):

English:	the newcomers **'captured'** a subset of the tRNAs (Freeland & Hurst 2004:91)
German:	die Neulinge . . . einige tRNAs und zugehörige Codons **»kidnappten«** (Freeland & Hurst 2004:93) [the newcomers . . . **'kidnapped'** some tRNAs and the related codons]

Here the translator in effect homes in on a particular kind of capture, and one with which readers will be familiar through numerous news reports, for example.

b) Explicitation (one example):

English: Oligodendrocyte glia produce myelin that **insulates** axons. (Fields 2004:57)
French: Les oligodendrocytes, d'autres cellules gliales, produisent la myéline qui **isole électriquement** les axones. (Fields 2004:59)
[The oligodendrocytes, another type of glial cell, produce the myelin that **electrically insulates** the axons.]

In this version, the translator chooses to add a supplementary detail—which is a very commonplace translation procedure in general, of course, if surprisingly rare in the set of data I am looking at.

c) Shift from non-rich image to rich image in conjunction with a change of mapping (6 examples):

English: how exactly would glia be affected by what they **heard**? (Fields 2004:56)
Russian: как влияют на работу глиальных клеток «**подслушанные**» ими нервные сообщения? (Fields 2004:25)
[how do the neural communications that they '**overhear**' influence the work of glial cells?]

In this case, the increase in specificity is concomitant with a change of mapping from BRAIN CELLS ARE SENTIENT to the far more vivid and specific BRAIN CELLS ARE EAVESDROPPERS.

Needless to say, all three of these shifts entail the addition of detail and/or specificity to the metaphorical expression.

iii) Third, there are a number of observations that are worth making about precisely what can occur when a rich image is not 'reduced' to a non-rich image:

a) It is not necessarily the case that one-word or otherwise 'simple' rich images are generally retained. In some instances, this is indeed what happens:

English: Two days later Schwann glia cells had formed **a bridge across the divide.** (Fields 2004:61)
Italian: Due giorni dopo le cellule gliali di Schwann avevano formato **un ponte attraverso la recisione.** (Fields 2004:55)
[Two days later the Schwann glial cells had formed **a bridge over the cut.**]

The key lexical item in this metaphor, *bridge*, is also translated 'straight' in the other four relevant target languages, and this is a pattern that also occurs in a number of other examples. On the other hand, in a range of other cases, this does not occur. In the following instances,

English: the brain's chemistry and **architecture** (Nestler & Malenka 2004:80)
English: **rewire** cells to glow in the presence of a particular toxin (Gibbs 2004:75)

the keyword in the original English expression—and sometimes the entire expression—is removed or omitted in every target language. As before (see Interlude One and Section 6.1.3.1.3 point ix)), there is no one unequivocal reason as to why this should occur, although, on the other hand, in the case of 'rewiring', there is limited evidence that the decision to move away from this particular metaphor may not be entirely random, since in the other similar example—quoted this time in full—the same thing occurs:

English: It would be the difference between introducing a single typo and **rewiring** the entire typewriter keyboard. (Freeland & Hurst 2004:87)
German: Der Unterschied entspräche dem zwischen einem Druckfehler und der kompletten Neuverteilung der Buchstaben auf der Tastatur einer Schreibmaschine. (Freeland & Hurst 2004:89)
[The difference would be equivalent to that between a typo and the complete redistribution of the letters on the typewriter keyboard.]
Russian: Есть большая разница между появлением одной-единственной опечатки и **поломкой** всей клавиатуры печатающего устройства. (Freeland & Hurst 2004:57)
[There is a great difference between the appearance of a single typo and the **breaking** of the entire keyboard of a typing device]
Polish: To tak, jakby porównać efekt jednej literówki do skutków zmiany układu klawiatury. (Freeland & Hurst 2004:59)
[It is like comparing the effect of a single typo to the result of changing the keyboard layout.]
French: cela introduirait la même différence qu'entre une faute typographique et le réaménagement complet d'un clavier d'ordinateur. (Freeland & Hurst 2004:62)
[that would introduce the same difference as that between a typing error and the complete rearrangement of a computer keyboard.]
Italian: È un po' la differenza che passa tra inserire un singolo refuso in un testo e riconfigurare tutta la tastiera di una macchina per scrivere. (Freeland & Hurst 2004:91)
[It is a bit like the difference that exists between inserting a single typo in a text and reconfiguring the whole typewriter keyboard.]

Chinese: 如果打個比方來看. . .就相當於打錯一個字母與重整鍵盤的字母位置之間的差別。(Freeland & Hurst 2004:81)
[If one were to use an analogy . . . it's equivalent to the difference between mistyping a single letter and reorganising the positions of the letters on the keyboard.]

Clearly, as part of a more detailed metaphor that occupies an entire sentence, there is scope for rewording this individual lexical item while retaining the overall vivid metaphorical force. On the other hand, a clear possible factor in the universal rewording that we see in this example is the fact that none of the target languages have a single, one-word translation equivalent for 'rewire' (although clearly in each case this is a meaning that frequently needs to be expressed by other means).

On the other hand, this is not so in the case of 'architecture', where a simple one-word equivalent does exist in each target language. This highlights what is to my mind one of the more noteworthy features of the data that I have looked at, which is the apparent lack of motivation for many of the instances in which a metaphorical expression is removed; for example, as far as I can see, there should be no reason why the lack of a one-word equivalent should trigger a rewording in six independently produced translations as we have seen in this last example. I hypothesise that the causes for this are complex and therefore not easy to predict, while if I were to speculate, I would suggest that some translators may worry that a particular metaphorical expression, which may be perfectly comprehensible to them in the source text, may for some reason not entirely 'work' in the target language and so take the decision not to let a simple direct translation stand. (The other matter that I also find somewhat difficult to explain is the seeming unpredictability of the solutions used to translate expressions on different occasions, as mentioned on a number of occasions earlier.)

b) It is probably also worth pointing out that a rich image can be retained even when the wording is reduced:

English: **Like a wave of cheering fans sweeping across a stadium**, the calcium waves spread throughout the entire population of astrocytes. (Fields 2004:60)

French: telles des **olas moléculaires**, les vagues de calcium se propagent dans les astrocytes. (Fields 2004:61)
[in the same way as the **molecular Mexican waves**, the waves of calcium spread in the astrocytes.]

(One further example of this has been found.) In other words, a particular rich image is (obviously enough) largely independent of the precise wording in which it is expressed, which may vary considerably between the source and target language.

iv) Finally, there are a number of points that can be made regarding non-rich images that are not shifted into rich images. For this, around 20–25 English examples and their translations were analysed to obtain an impression of the different approaches used, which means that this list should by no means be considered in any way exhaustive.

a) The 'straight' translation approach described in Section 6.1.3.1.3 point i) can be seen in 53 expressions:

English: neuroscientists focused their work narrowly and left the **big picture** alone. (Solms 2004:84)

Polish: badacze mózgu skoncentrowali się na wycinkowych zagadnieniach, tracąc zainteresowanie **ogólnym obrazem.** (Solms 2004:76)
[researchers into the brain concentrated on partial problems and lost their interest in the **general picture.**]

Most of these involve single-word metaphors (e.g. *plasticity*, *overshadowed*, *reward*), although it would be incorrect to think that all such metaphorical expressions were handled in this manner by translators. None of the examples entail a change of mapping, although in some other instances this is precisely what occurs, as the following example illustrates.

b) The second procedure also entails a relatively straight translation, but in this case with a change of mapping:

English: biological approaches to mental illness gradually **overshadowed** psychoanalysis. (Solms 2004:84)

German: biologische Erklärungen für psychische Krankheiten **verdrängten** die psychoanalytischen Ansätze
[biological explanations for mental illnesses **displaced** the psychoanalytical approaches]

Russian: биологический подход к изучению психических заболеваний мало-помалу начал **вытеснять** психоанализ.
[the biological approach to the study of mental illnesses gradually started to **force out** psychoanalysis.]

French: les approches biologiques de la maladie mentale **supplantèrent** peu à peu l'approche psychanalytique
[the biological approaches to mental illnesses **supplanted** little by little the psychoanalytical approach]

Six such examples have been identified. Here the SURPASSING IS OVERSHADOWING of the English sentence gives way in the German to SURPASSING IS SUPPRESSING and in the Russian and French to SURPASSING IS FORCING OUT.

c) Thirdly, from the discussions presented in Sections 6.1.1.3 and 6.1.2.3 it should be clear that there is a strong tendency for expressions based on non-rich images to consist of no more than one or two words. However, there is no absolute link between the number of words a metaphorical expression contains and its level of richness, and occasionally the size of an expression increases in translation without it becoming significantly richer. This can happen, for example, when a source-language term is translated according to sense:

English: chemical **messenger** molecules (Fields 2004:55)
Italian: molecole chimiche **che trasportano un messaggio** (Fields 2004:50)
[chemical molecules **that transport a message**]

I have identified a total of three examples of this. While a straight translation would have been possible the translator here seems to have deemed it necessary to explicitate or explain what she considered to be a potentially obscure expression for the readers. In the process, however, the expression remains a non-rich image.

Unlike the first, the second example is non-terminological in nature:

English: oligodendrocyte glia cells **wrap around** axons in the central nervous system (brain and spinal cord). (Fields 2004:56)
Russian: а глиальные клетки другого типа (олигодендроциты) **образуют оболочки вокруг** аксонов в центральной нервной системе (т.е. в головном и спинном мозге). (Fields 2004:24)
[glial cells of the other type (oligodendrocytes) **form coverings around** the axons in the central nervous system (i.e. the brain and spinal cord).]

This tendency has been observed in five examples. Even though the Russian translator avoids the informality of the original, strict terminology is not employed. However, once again, the expression in the target text remains a non-rich image.

These appear to be the main procedures involved in translating metaphorical expressions when seen from the point of view of whether they are rich or non-rich images. A number of less common procedures that entailed leaving non-rich images as non-rich images may well emerge if more data were to be analysed.

6.1.3.2.3 METAPHORICAL EXPRESSIONS THAT HAVE BEEN ADDED IN TRANSLATION

It is worth noting that of the 14 metaphorical expressions that have been identified as existing in one or more of the translations, but not in the source text, 4 are rich images and 10 non-rich images. The four rich image expressions involve the German morpheme *Erb-* 'heredity' that was discussed in Interlude Two. Based on this very limited information, it seems that it may be the case that added rich images are not only few in number but also none of them is particularly bold or creative.

6.1.3.2.4 CULTURALLY EMBEDDED EXPRESSIONS

There appears, not surprisingly, to be a clear correlation between an expression's richness and the extent to which it is embedded in a particular culture. It is worth noting that bold source-language, rich-image metaphorical expressions that appear to be embedded in the source culture are sometimes (but not always) retained:

English:	tiny cellular **butlers** . . . fetch the specified amino acids (Freeland & Hurst 2004:86)
French:	de minuscules **majordomes** cellulaires . . . assemblent les acides aminés adéquats. (Freeland & Hurst 2004:61) [miniscule cellular **butlers** . . . assemble the appropriate amino acids]
Italian:	minuscoli «**maggiordomi**» cellulari . . . vanno a cercare gli amminoacidi specificati (Freeland & Hurst 2004:90) [miniscule cellular **butlers** . . . go in search of the specified amino acids]

This expression is, on the other hand, omitted in German and Chinese, removed in Russian and modified to 'couriers' in Polish, even though appropriate equivalents exist in all those languages.

Of the 38 target-language examples examined, in 29 cases, the metaphorical expression was retained (though sometimes in modified form). Besides these, seven metaphorical expressions were removed and nine omitted.

6.1.3.2.5 REVISITING AL-HARRASI'S PROCEDURES

We are now in a position to appraise Al-Harrasi's use of rich images and image schemas in his list of translation procedures. As described in Section 3.2.2.5, his taxonomy of 11 procedures contains a total of 3 that involve the interplay between rich images and image schemas:

1.4. Same Image Schema and Rich Image Domains
1.6. Using a Different Rich Image That Realises the Same Image Schema Realised by the Rich Image in the Source Text
1.7. From the Rich Image Metaphor to Image-Schematic Representation (2001:277–88)

A fourth, 1.5. 'Same Rich Image Metaphor but Alerting the Reader to the Mapping', also refers to rich images but not in combination with image schemas.

Al-Harrasi's proposal is a highly interesting and original one. The problem for us, however, is that in our data there are only 10 English expressions out of a total of 268 that can be easily attributed to one or other image schema, while as discussed in Section 6.1.3.1.2, in each of these 10 cases, this apparent ultimate derivation from an image schema is overwritten by layers of propositional knowledge, which means that, on the surface, there is effectively very little that distinguishes these metaphorical expressions from all the others.

Let us first of all look at one of Al-Harrasi's own examples, albeit one that exemplifies another of his procedures, '1.2. Concretising an Image-Schematic Metaphor':

Arabic: . . . والغاء الغبن والحيف الذي كان قد اصاب العراق **في صميم** كيانه قبل يوم النداء . . .

[to cancel the injustice and unfairness which had hit Iraq **in the core of** its entity before the day of call [i.e. the day of Iraq's invasion of Kuwait]] (from a speech delivered by Saddam Hussein)

English: to cancel the injustice and unfairness that had hit Iraq **in the heart of** its entity before the day of call (2001:279)

Al-Harrasi attributes this to what he terms the CENTRE image schema. Here we see the translator concretising the image-schematic metaphorical expression by choosing to translate 'core' with 'heart', which possesses far deeper resonances in the context of human experience. Although 'heart' is not identified by Al-Harrasi as a rich image, this is in line with the procedure discussed in Section 6.1.3.2.2, point ii) a).

The list of image schemas that Al-Harrasi uses seems to be different from the one that I presented in Section 6.1.1.1, as it includes additional items such as WEIGHT/BALANCE, ESSENCE/CORE and DISGUISE/REVELATION (2001: 131–58) (and, indeed, CENTRE). In addition, if we look at a second example of his—relating to his procedure 1.7. 'From the Rich Image Metaphor to Image-Schematic Representation'—we find that it is attributed to an image schema that Al-Harrasi identifies as STAGNATION:

Arabic: واذا كان قد مر على هذه المدينة حين من الدهر **كاد يطمرها غباره**

[if in the past passed on this city period of time **whose dust was about to bury it**, . . .] (from a speech by Sultan Qaboos of Oman)

English: Notwithstanding the **era of stagnation** in the past, . . . (2001:285)

Al-Harrasi correctly states that, unlike the source-text phrase that it replaces, the notion of *stagnation* is not specifically linked to the geographical area that is being talked about, but is more general in nature (2001:285), which is why this instance of translation is categorised in the manner that it is. However, referring back to our Section 6.1.1.1 discussion of image schemas, it is possible that the concept of *stagnation* may be stretching the notion of 'basic cognitive structure' too far. In other words, within the terms of this present work at least, the English translation of this example cannot in fact be thought of as relating to an image schema.

On the other hand, the contrast that is frequently seen between rich images and non-rich images that was discussed at length earlier is in many ways parallel to that between rich images and image schemas; in each case, on the one hand, we have a vivid, detailed image that creates complex, contextualised associative patterns in the mind of the reader, and on the other, a generalised, non-specific expression with associative patterns that are considerably less well developed.

For this reason, I consider Al-Harrasi's examples to be directly parallel with many of those drawn from my data that clearly demonstrate similar shifts in translation to that illustrated by the 'stagnation' example, with a move from a more to a less highly specified image taking place in each case, although on each occasion it can be understood without reference to an image schema.

It should be pointed out that Al-Harrasi's scheme of translation procedures listed in Section 3.2.2.5 does not allow for the eventuality of a straight image-schematic metaphor being replaced by a rich image in translation. This is presumably because his data contains no such examples. Such instances—or ones closely parallel to them—do, however, exist as we see from the following:

English: the dopamine **latches** onto its receptor (Nestler & Malenka 2004:82)

German: [das Dopamin] an ein passendes Rezeptormolekül **andockt** (Nestler & Malenka 2004:38)
[[the dopamine] **docks** with a suitable receptor molecule]

This example, which exemplifies the procedure described in Section 6.1.3.2.2, point ii) a), demonstrates a clear movement from the less specified to the more specified—a shift that should definitely be taken into account in an exhaustive model.

In the light of this discussion, it should first of all be reiterated that Al-Harrasi's insights are not completely applicable here, not least because, unlike his data, mine includes very few image schema-based metaphorical expressions. However, given that the borders between image schema and non-image schema are not precisely defined in his thesis, I believe it would be quite in the spirit of his original proposals to generalise them by not specifically invoking the concept of the image schema. In this way, his fundamental notion of an interplay between the more and the less vivid and/or specified becomes a very powerful one in terms of its ability to explain the translation phenomena that are engendered by the presence of metaphorical expressions in texts.

6.1.3.2.6 CONCLUSION

What we can see is that there is a far greater tendency to replace a rich image with a non-rich image in translation than vice versa, and in line with this, loss of specificity is commoner than the opposite. However, taking into consideration the effects of other processes such as omission and removal as well, the overall result is that the ratio of rich images to non-rich images remains broadly unchanged in the translated texts at approximately 40:60, which is the original ratio stated in Section 6.1.3.2.1. In terms of expressions that are widely or universally omitted, removed or, on the other hand, retained, this 40:60 ratio broadly obtains in each case.

The section also broadly confirms the validity of Al-Harrasi's approach to analysing modifications to metaphorical structures brought about in translation, although it does propose a number of changes to the scheme that was originally developed by him.

6.2 Image Metaphors

Unlike the metaphor types that were considered in Section 6.1, image metaphors are not based on image schemas and propositional knowledge but rather reflect physical similarity. Consequently, their presence in a text depends on the extent to which shapes, textures and other physical phenomena form the subject matter. As discussed earlier, this type of metaphor is acknowledged by the cognitive theory of metaphor, but has not to date been looked into by metaphor scholars to any great degree, a situation that is reproduced to an even greater extent in the study of metaphor in translation.

6.2.1 Background

As explained earlier, image metaphors are one of the four basic types of metaphor described by Lakoff. This metaphor type entails 'the mapping not of concepts but of mental images from one source of knowledge onto the mental images from another' (Gibbs 1994:258). If image-schematic metaphors are based on how entities interact with each other physically, then image metaphors focus on physical similarity. Because of this, image metaphors can be thought of as 'resemblance-based' (Evans & Green 2006:293). A particularly famous (or, at least, widely quoted) example ('My wife . . . Whose waist is an hourglass') comes from a translation of a poem entitled *Union libre* by the early twentieth-century French surrealist poet André

Breton, and for a long time image metaphors were supposed to occur almost exclusively in literary texts; in addition, image metaphors have generally been held to be 'one-shot' in that the mapping 'sanctions the lexical extension of only one word' (Lakoff 1987c:195).

Elsewhere, Lakoff provides six criteria for distinguishing image metaphors from 'very general conceptual metaphors':

1. Image metaphors are not used over and over again.
2. They are not used in 'everyday reasoning'.
3. They do not underlie a 'system of words and idiomatic expressions'.
4. They map 'image structure instead of propositional structure'.
5. They do not interpret the abstract in terms of the concrete.
6. They have no basis in 'experience and commonplace knowledge' that determines 'what gets mapped onto what'.

(1987b:221)

Because of their basis in physical similarity, in the Breton poem cited earlier, *Whose waist is an hourglass* is considered an image metaphor, but the line that precedes it, *Whose thoughts are summer lightning*, is not, because of the absence of a target image (since 'thoughts' does not create a visual or other kind of image), which means that the mapping has to be deemed conventional in nature (1987b:221). Elsewhere, Lakoff states that a metaphorical expression can combine elements of image metaphor and proposition-based conceptual metaphor 'in cases where complex knowledge about the source image is carried over into the target domain' (1987c:195), although this insight was not expanded on for a number of years.

The fact that thinking on image metaphors remained largely static for a long period can be seen from the tendency on the part of many writers to reproduce this one same Breton example as a kind of 'standard' example of this type of metaphor. As far as it has been possible to ascertain, it was first quoted by Lakoff (1987b:219–22); subsequently, it has been referred to by many other writers (for example, Lakoff & Turner 1989:90; Lakoff 1993:229–30; Gibbs 1994:258–9, 1999a:32–3; Yu 1998:31; Kimmel 2002:52; Caballero 2003a:87–8, 2006:77; Evans & Green 2006:293–4; Kövecses 2006:129, 2010:57, 2010:327—Lakoff 1987b and Gibbs 1999a, incidentally, being possibly the only scholars to point out that the poem being commented on is actually a translation). In other words, throughout most of the period since 1987, this topic has not been the subject of a serious collective research effort, with a range of scholars simply proffering a standard account (and example) of what for them is a matter of peripheral interest.

Caballero stands out from the other names on the aformentioned list, as she challenges the main thrust of Lakoff's view with the findings of a research project that constitutes one of the only studies devoted to image metaphors of which I am aware (see Caballero 2002, 2003a, 2003b and

2006). Because of their supposed association with literary and poetic texts, Caballero argues that image metaphors have tended to be ignored by cognitive linguistics, which has usually focused its attention more on everyday language use (2003a:87–8), and that when they have been discussed at all, they have typically been characterised as possessing an 'ad-hoc or fleeting nature' (2003a:101).

On the basis of her study of 95 architectural reviews, she presents a large number of new examples of this type of metaphor, including the following (all of which preserve the emphasis of the original):

- the basic part of the bank building is *a three-sided doughnut* (2003a:89–90)
- the decision to air-condition lower-floor public spaces required ingenious *weaving of ductwork* in ceilings (2003a:90)
- the traditional *supermarket box* (2003a:93)
- the *pinwheel plan* (2003a:93)
- the *reptilian* green copper cladding (2003a:96)

One of Caballero's main contentions is the following:

> Image metaphors which, in principle, appear to rest upon a single or minimal correspondence of physical traits are nevertheless capable of triggering a rich set of inferences, yet this only happens when their linguistic realisations are seen within their context of occurrence.
> (2003a:102)

For this reason, the view of image metaphors as 'ad hoc or fleeting' is too simplistic (2003a:101); rather, they appear to be 'conventional devices underlying architects' conceptualisation of their specific world' (2003a:101). In the texts that Caballero studied, image metaphors appear to provide 'an essential heuristic tool' (2003a:87) for communicating discipline-related phenomena. On the basis of her data, she argues against the 'clear-cut distinction between visual and conceptual knowledge', as well as that between the 'ordinary' and the 'one-shot' metaphor (2003a:88), and also identifies a crossover category consisting of conceptual metaphors that appear to be 'tinted with imagistic overtones as well' (2003a:89). Interestingly, in an entirely different subject area, many of the English, French and Spanish medical terms that are listed by Divasson and León (2006:59–61)—including a large number that relate to pathology and morbid anatomy—are also image metaphors. Examples include *vertebral column*, *nerve root*, *moon face* and *raspberry tongue*, with corresponding phrases in French and Spanish. As with the architectural examples studied by Caballero, such terms tend to occur in types of discourse that focus on physical objects (or, more rarely, sounds, tastes and so on).

6.2.2 *How This Concept Features in the Data*

Since the presence of image metaphors in a text is very largely a function of the subject matter, the more physical shape is described and discussed, the greater the number of image metaphors we can expect to see. Of the 21 image metaphors identified in the data (see Table 6.1), few contain a significant element that is bound by culture-related considerations, based as they are on physical similarity. Consequently, one of the questions that will be investigated in Section 6.2.3 will be whether such expressions are any easier to translate than other more culturally embedded ones, to the extent to which it is possible to distinguish between them.

Besides the physical similarity requirement that all image metaphors must meet, it should be noted that many of the mappings link entities that are vastly different in terms of size; for example, neurons are seen as trees (13 examples) and a synapse as a bridge (2 examples). This means that, unlike the architectural descriptions that Caballero analyses, the expressions in the present corpus frequently refer to entities that, if not abstract, cannot be directly perceived. Very much along the lines of Caballero's work, the data collected appears then to contradict a number of the criteria for identifying image metaphors set out by Lakoff in Section 6.2.1. Taking the mapping NEURONS ARE TREES, which gives rise to a whole range of different usages, we can see that criteria 1 and 3 do not apply, nor do 2 or 5, at least to some extent. In other words, the examples contained in my data call the theoretical basis of how image metaphors should be understood seriously into question.

On the other hand, Lakoff's undeveloped comment, also cited in Section 6.2.1, that image-based and proposition-based elements can be combined in a single metaphorical expression 'in cases where complex knowledge about the source image is carried over into the target domain' (1987c:195) does find considerable confirmation. This is precisely what we see happening in the mapping NEURONS ARE TREES, for example. Other such 'transitional' examples—where it is unclear whether the mapping is on the basis of physical similarities or conceptual parallels—include *as animals move up the evolutionary ladder* (Fields 2004:61) and the two examples that involve the notion of a *network*, in which the concepts in question are both imagistic and propositional in nature. Each of these examples is categorised on the basis of which of the two elements is felt to predominate.

One further observation is that the data reveals that the identification of image metaphors with one-shot metaphors is probably inappropriate, as there are a range of conceptual mappings that are associated with the use of a single lexical item and a number of image metaphors that possess a significant conceptual element, as has just been demonstrated. Finally, there is an important correlation between rich images and image metaphors, although while the majority of image metaphors in the English data are rich images, only a small number of rich images are image metaphors, which is a pattern that is mirrored in the other languages represented in the data.

6.2.3 Translation Procedures and Analysis of Examples

As already hinted, I believe that the discussion that follows is the first translation studies investigation into image metaphors. The 21 English image metaphor examples contained in the data consist of 13 relating to a single mapping, NEURONS ARE TREES, and 8 miscellaneous ones that draw on source domains that range from fire and bridges to Mexican waves and Christmas lights. All in all these English expressions give rise to 114 translated examples. As was the case with the image-schematic expressions, all image metaphors belong to only two existence categories, most of the miscellaneous expressions being examples of concretisation, and all the instances of NEURONS ARE TREES exemplifying plantification. Interestingly, as illustrated by Table 6.2, the data contains no shifts from image metaphor to another type of metaphorical expression. In addition, the proportion of expressions removed is about half the average for the data as a whole (around 1 in 10 as against 1 in 5). None of the examples entail a change of existence category; there is only one instance of repurposing, while three out of the four modifications in conventionality are in the direction of conventionalised to innovative (which is in line with the rest of the data). I found no new image metaphors in any target language. On the basis of all these facts, the tentative conclusion that can be drawn is that image metaphors do not pose any special problems for translators.

The following procedures are worthy of note, although they are very tentative because of the small number of examples available for each.

i) First, a number of image metaphor expressions are retained essentially unchanged:

English: THE MENTAL PICTURE most people have of our nervous system resembles **a tangle of wires that connect neurons.** (Fields 2004:55)

Russian: Мы представляем себе нервную систему в виде **переплетения проводов, соединяющих нейроны.** (Fields 2004:24)
[We imagine the nervous system in the form of **a tangle of wires connecting neurons.**]

Polish: Układ nerwowy w rozumieniu większości ludzi to **plątanina przewodów łączących neurony.** (Fields 2004:28)
[The nervous system in most people's understanding is **a tangle of wires connecting neurons.**]

French: On se représente souvent le cerveau comme **un enchevêtrement de câbles connectant les neurones.** (Fields 2004:56–7)
[One often imagines the brain as **a tangle of cables connecting neurons.**]

Italian: L'immagine mentale più diffusa del sistema nervoso è quella di **un groviglio di cavi che collegano fra loro i neuroni.** (Fields 2004:50)

[The most widespread mental image of the nervous system is that of **a tangle of cables that connect the neurons together.**]

Chinese: 大部份的人會把我們的神經系統想成是**一堆連結著神經元的糾纏線路**。(Fields 2004:35–6)

[Most people are likely to think of our nervous system as **a heap of tangled circuits that connect neurons.**]

I have identified 45 of these in the data. In each case, whatever minor rewordings there may be do not affect the metaphorical expression in any significant way, and the mapping is preserved unmodified.

ii) Second, a total of three expressions are made more implicit in translation:

English: **Like a wave of cheering fans sweeping across a stadium,** the calcium waves spread throughout the entire population of astrocytes. (Fields 2004:60)

French: telles des **olas moléculaires,** les vagues de calcium se propagent dans les astrocytes. (Fields 2004:61)

[in the same way as the **molecular Mexican waves,** the waves of calcium spread in the astrocytes.]

In this instance, a description is replaced by a precise term. Implicitation can assume a number of different forms, however, and the other examples involve the abbreviation of a term and the removal of a cultural reference respectively.

iii) The third procedure is the opposite of the previous one—that is to say, explicitation:

English: Each neuron has a long, outstretched branch**—an axon** (Fields 2004:55)

Polish: Od każdego neuronu odchodzi długa wypustka, najczęściej rozgałęziona na końcu, zwana **wypustką osiową, neurytem lub aksonem.** (Fields 2004:28)

[From each neuron there emerges a long projection, usually branching at the end, known as **the axial projection, neurite or axon.**]

In each of the 5 examples of this procedure, while the translator's motivation can only be guessed at, further wording is added.

iv) Fourthly, there are 12 examples of an expression becoming more abstract:

English: Each neuron has **a long, outstretched branch** (Fields 2004:55)

Russian: Каждый нейрон снабжен **одним длинным отростком** (Fields 2004:24)

[Each neuron is equipped with **a single long shoot**]

Polish:	Od każdego neuronu odchodzi **długa wypustka, najczęściej rozgałęziona na końcu** (Fields 2004:28) [From each neuron there emerges **a long projection, usually branching at the end**]
French:	Chaque neurone a un long <u>prolongement</u> (Fields 2004:57) [Each neuron has a long <u>extension</u>]
Italian:	Ogni neurone ha **una lunga ramificazione molto estesa** (Fields 2004:50) [Each neuron has **a long, very extended ramification**]

As can be seen, this example relates to the mapping NEURONS ARE TREES. Three of the abstraction shifts leave the mapping unchanged, but in the case of the French one, the metaphorical component is removed entirely—a phenomenon that is not unique in the set of examples collected.

v) The fifth procedure entails a change to a different image, as represented by two examples:

English:	**Like a wave of cheering fans sweeping across a stadium**, the calcium waves spread throughout the entire population of astrocytes. (Fields 2004:60)
Italian:	**Proprio come un'ondata di tifosi che si riversa in campo al termine di una partita**, anche i flussi di ioni calcio si diffondono invadendo l'intera popolazione di astrociti. (Fields 2004:55) **[Just like a wave of fans who pour onto the pitch at the end of a match**, the flows of calcium ions spread out to invade the entire population of astrocytes.]

Here, in addition to a different image being used, the mapping is also changed (NEUROTRANSMITTERS ARE MEXICAN WAVES becoming NEUROTRANSMITTERS ARE WAVES OF FANS). It should be pointed out that there appears to be a possibility that in this example, the change comes about as a result of the original English not having been properly understood. A further, more incidental observation is that the use of the word *calcio* constitutes a pun, as this item can mean either 'calcium' or 'football'.

In the following example, on the other hand, the shift appears to be more unequivocally intentional:

English:	forming a cell-free void that would **act like a highway separating burning forests on either side.** (Fields 2004:58)
Russian:	на две части, оставив между ними **зазор**, не содержавший клеток и разделявший популяцию астроцитов. [in two parts, leaving between them a **forest clearance gap** that contains no cells and separates the population of astrocytes.]

Here the mapping changes from BRAIN ACTIVITY IS FIRE to THE BRAIN IS A FOREST, although the target expression is still an image metaphor.

vi) The sixth and final procedure is cultural reinterpretation, as seen in this single example:

English:	the glia suddenly began **to light up like bulbs on a string of Christmas lights** (Fields 2004:57)
Russian:	они, **подобно елочным лампочкам**, начали загораться. [they, **like lights on a fir-tree**, started to light up]

It should be pointed out that this particular example is more complex, as it combines cultural reinterpretation (i.e. the association of the fir-tree with the New Year rather than with Christmas, which continues to predominate in Russia) with implicitation.

Finally, it is perhaps worth pointing out that there are no instances of metaphorical expressions that are added in translation (they are all in fact based on propositional knowledge).

Overall, image metaphors represent a relatively stable category in translation, and as mentioned earlier, there are no examples of an expression being shifted into or out of it. Besides the large number of image metaphors that are translated without any significant modification (procedure i)), the tendency to become more abstract (procedure iv)) is also relatively prominent. Of the procedures identified, only i) and v) reflect features that are specific to image metaphors, while all the others are more general in nature.

6.3 Conclusion

While image metaphors were simple enough to classify, the same could not be said for image-schematic expressions, as it proved very difficult to disentangle these items, which were invariably supplemented by elements of propositional knowledge, from expressions that were derived purely from the latter. Regarding the hypothesis advanced in Section 6.1.2.1 to the effect that image-schematic metaphors tend to present the translator with fewer problems by dint of their universality, it seems not to be borne out by the data, not least, however, because of the layers of propositional or cultural knowledge placed on top of the image-schematic base, which it could be argued 'contaminate' the image-schematic purity of these expressions and thus render the comparison unhelpful. On the basis of my data, in fact, it would appear that it is image metaphors that produce the fewest problems for translators. However, the hybrid nature of all the image-schematic expressions that I have looked at bears out Al-Harrasi's original model and indicates that his approach is indeed a fruitful one.

At the same time it should, however, be pointed out that the notion of propositional knowledge represents an important addition to Al-Harrasi's model as the concept of the image schema may have been over-generalised in the framework that he proposes. Furthermore, it should be borne in mind that the corpus contains no examples of his two procedures 1.5 'Same Rich Image Metaphor but Alerting the Reader to the Mapping' and 1.7 'From the Rich Image Metaphor to Image-Schematic Representation'.

What is perhaps striking—but at the same time not surprising—about the propositional knowledge-based expressions is the sheer number of procedures that have been identified. That said, I do not in fact believe that the listing presented earlier is inevitably exhaustive, as analysing a larger sample may well yield further minor procedures. As it is, this extensive listing can probably be seen as the most generalisable single procedure list, not specific to any particular parameter, that this research has produced.

Note

1. Passages from the book, most of which fall within this section, constitute an amended version of the following article: 'Scientific Rich Images in Translation: A Multilingual Study', in *JoSTrans, The Journal of Specialised Translation* (2014) 21: 35–51.

7 Conclusion

The research presented in this book has been relatively complex, not only drawing on at least two different disciplines but also encompassing six target languages, a wide range of types of metaphor, several different scientific areas, six parameters of analysis and the output of numerous translators. The following pages set out some of the conclusions that can be drawn and touch on three of the book's major research focuses. After this, the chapter will close with some suggestions as to how research into this area can be taken forward.

7.1 Conclusions from the Research

The three areas that this section comments on are the relationship between translation studies and metaphor research, changes that occur on the metaphor level during translation and the development of a wide-ranging list of translation procedures specific to metaphor.

7.1.1 Grounding Research into Metaphor in Translation in Metaphor Research

In Chapter Three, the conceptual metaphor theory as envisaged by Lakoff and Johnson was presented as a framework that is not perhaps ideal for translation studies research. This is partly because of its downplaying of the extent of interlingual and intercultural variation in metaphor, and because it does not recognise the importance of basing arguments on authentic rather than made-up, idealised examples. On the other hand, the way in which this theory has been implemented by a large number of other metaphor scholars, who recognise the importance of language and culture and analyse metaphor in authentic texts, provides a much more promising starting point for research into metaphor in translation, and it is this approach that I have used in large parts of my research. A further and more minor point is that the de facto convention in cognitive linguistics of expressing metaphorical mappings in the A IS B format is unnecessarily restricting when it comes to verbalising complex scientific source-target relationships.

Regarding Al-Harrasi's suggested list of procedures, this study confirms the importance of his insight that the interplay between the more and the less vivid and/or specified is one of considerable explanatory power. On the other hand, it argues that as it stands, it is not complete, and it would be further strengthened by increasing its generalising power—for example, by adding the concept of propositional knowledge. In addition, it may be desirable to revisit some of the specific procedures that he proposes.

Another more general finding of the research is that the theoretical frameworks used for classification purposes differ in terms of their explanatory usefulness. In the case of concepts deriving from the conceptual metaphor theory, for example, it appears that image-schematic metaphors are treated broadly in the same way as image metaphors and propositional knowledge-based expressions, although this is perhaps understandable in view of the problem with image-schematic expressions discussed earlier. With existence categories, on the other hand, it seems that translators may display certain tendencies in the way in which they distinguish between expressions belonging to different classes. Furthermore, in the case of richness, a greater tendency to move from rich image to non-rich image than in the opposite direction can be observed. Next, the basic divide between purpose categories that are central or peripheral to the aim of communicating scientific ideas will influence the likelihood of certain procedures being opted for. Finally, the extent to which a metaphorical expression is bold or creative will apparently influence the ease with which it can be translated. In general terms, I therefore believe that the validity of the categories and distinctions represented by the six parameters has been at least partially confirmed. In addition, the usefulness of at least some of them has been demonstrated through the way they were applied to refining Al-Harrasi's list of procedures, for example. That said, however, because of its breadth of coverage the research reported on in this book should be regarded as in some ways exploratory, and further work focusing in greater detail on particular aspects of it is certainly needed in order to develop a more nuanced understanding of some of its contentions.

Much has been taken from metaphor research in the course of this investigation, so it is now necessary to consider what may be offered back to that discipline in return. The answer to this is perhaps threefold: firstly, a range of frameworks for analysing metaphorical expressions; secondly, a number of individual insights; and thirdly, a possible extension to the discipline's applied wing.

Regarding the first point, the frameworks are those from which the theoretical ideas have been drawn for the purposes of categorising metaphorical expressions. While it might potentially be of use to metaphor scholars, it should be noted that the particular values selected for some of the parameters (for example, mapping, existence category and purpose) are specific to the corpus and may need to be adjusted if they are to be used with non-scientific types of data. In other cases, such as image-schematic versus propositional knowledge-based metaphorical expressions, the proportionalities may prove to be very different if applied to other subject areas or types of text.

Secondly, many specific insights can be obtained by drilling down into the monolingual English data. Some of the more significant ones are as follows. One of the most interesting concerns image-schematic metaphorical expressions, as it transpires that complex examples of this type of metaphor, which is considered to be of great importance in conceptual metaphor theory, are in fact very rare, in the present data at any rate. The overwhelming proportion of metaphorical expressions—around 88.4%—are in fact based on propositional knowledge, while the few image-schematic expressions that I have considered to be of sufficient interest in the context of translation studies research to be worth recording (around 3.7% of the total) are themselves overlaid with layers of propositional or cultural knowledge. Likewise, image metaphors, another category of metaphor that is downplayed in much metaphor research, are relatively more frequent at around 7.8%. Indeed, although not as detailed as the studies of this area conducted by Caballero, the relevant sections of the book represent a modest new contribution to the sparse research literature on this topic. In this respect, one insight that the book achieves is that the term 'one-shot metaphor' is not in fact coterminous with 'image metaphor'. The situation with rich images is similar: once again not accorded significant theoretical attention, this category of metaphorical expression occurs in an approximate ratio of 40:60 as against non-rich images, which means that in this respect, too, the book offers a brief case study on an under-researched topic that is also underplayed in terms of theory.

Thirdly—and much more generally—works on metaphor in translation should be recognised as belonging to the applied branch of metaphor research as well as translation studies. In the event that the different areas of metaphor research were to be formally mapped out along the lines of Holmes (2004), metaphor in translation certainly deserves to be included alongside other topics that have hitherto received greater recognition within the discipline.

The final point to be made in this section is that the study has not proved particularly useful as a means of testing the validity of metaphor research categories and concepts—at least not in the sense of probing (by implication) their psychological reality in the minds of translators. It had been specifically hoped, for example, that it would be possible to see if image-schematic expressions were treated differently by translators because of their supposed 'universal' nature, but in the event, this proved not to be feasible—not least because of the very small numbers of such expressions in the corpus.

7.1.2 *Changes to Single Metaphorical Expressions and Entire Underlying Structures in Translation as an Aspect of Scientific Communication*

In terms of my data's fit with the two 'general-purpose' schemes of procedures presented in Section 3.2.2.5, instances of each of Pedersen's six baseline categories are not hard to find. The same is almost true of Pym's seven solutions, except that I was unable to identify any translated expressions that might be thought of as exemplifying perspective change. This could be because this solution does not need to be brought into play for rendering metaphorical language or else it might point to an inadequacy in my analysis. In the case of most or all of the other solutions, some but not all of Pym's suggested subcategories can be located in my data.

Overall, there appears to be no clear consistency in terms of how individual metaphorical expressions are translated into the six languages. Indeed, one is almost tempted to speak of 'unpredictability': not only do the translators working in the different target languages frequently use different procedures, but individual translators are not always consistent in the way in which they repeatedly translate particular metaphorical items. (This was one of two features that I found to be difficult to rationalise, as described back in Section 6.1.3.2.2.) Restating this in slightly more theoretical terms, it does not seem to be the case that an expression belonging to a particular category will with any reliability give rise to the use of a certain procedure. There are a small number of metaphorical expressions that are universally retained or deleted, but these expressions are greatly outnumbered by others that receive completely disparate treatments in the various target languages, and in general, there appears to be no way of predicting what an expression's likely fate will be either overall or in a particular language. As far as is possible to tell on the basis of the data, it appears that the translations are less metaphorical than the original texts. This conclusion is based on the very large numbers of expressions that are removed or omitted in translation—a tendency that is probably not counterbalanced by the presence of expressions that are added in translation. (The lack of obvious motivation for such removals is the other 'difficult feature' that I mentioned back in the same section.)

Chapters Four to Six, the three main content chapters, investigated in detail how different metaphors and/or metaphorical expressions are handled in translation depending on their characteristics. Detailed conclusions were reported at the end of each section. All six parameters are considered to be at least potentially fruitful and deserving of further investigation. No one parameter suggests itself as the ultimate key that unlocks the mysteries of translator behaviour in this area. However, it is possible to correlate translators' decisions with different metaphor types in the case of existence category, purpose, conventionality and richness to a greater extent than with mapping and provenance. In this connection, it emerges that there is no clear tendency to treat the different instantiations of a particular mapping in a similar manner, while image-schematic expressions appear to be neither easier nor harder to translate than other types.

The Holy Grail of significant far-reaching evidence that translators intervene at a high level to modify metaphorical structures to fit in with the requirements of their language and culture has not been located. Indeed, it is not generally possible to observe any kind of coherent strategy being used by translators in order to create new metaphorical systems in any target texts; the source-language mappings are generally conveyed, usually in a weakened form, or with certain modifications, by virtue of a proportion of the associated metaphorical expressions being rendered. There is an inevitable degree of scattering, which leads to the creation of many new mappings, most of which have only one example associated with them.

Regarding the question of whether or not we need to construct special laws for metaphor translation, the fact is that it is possible to conceptualise what happens to metaphor in translation in two different ways: the specifics of the analysis (for example, what happens to rich images in translation) are peculiar to metaphor in translation but at the same time there appears to be a wide-ranging higher-level compatibility with broader-based translation studies observations—a compatibility that is reflected, for example, in the mixed general/specific nature of some of the lists of procedures presented in the book. It remains my view, however, that there is no point in creating what might be regarded as a sub-discipline (or autonomous sub-theory) to explain metaphor-related translation phenomena if they can be subsumed within the existing frameworks of descriptive and theoretical translation studies. In this way, the work confirms the position that was adopted on this matter in Section 3.2.2.7.

The decision to make this a multilingual investigation has had a significant impact on the nature of the project. Certain lines of investigation have been rendered impracticable (looking at cultural and other contextual factors concerning a particular target language, for example), while at the same time, the pursuit of other questions has been enabled. There are a number of advantages that have been gained through the use of a multilingual approach. These include a gearing effect on the volume of data and the removal of the analysis from the immediacy of a single linguistic and cultural context. An additional advantage might be the possibility of bringing what is specific in the professional norms typical for translators working into a particular language into greater relief through the contrast that becomes available with those of other languages, although that was not an idea that was pursued in this project. The greatest benefit it has provided, however, is probably the new list of procedures presented in Section 7.1.3, which may well possess a greater authority than would otherwise be the case.

Also of interest is the fact that little evidence has been revealed to support the idea that particular source-text expressions engender specific solutions, or that what is perceived as difficult is consistent across translators working into all languages. Indeed, the question of difficulty remains somewhat problematic. On the one hand, given the high number of shifts, removals and omissions of individual metaphorical expressions it is impossible to argue that this is an area that is problem-free for the translator. On the other hand, translators generally appear to avoid addressing any possible high-level issues connected, for example, with the interlingual compatibility of mappings. This would indicate that translators give greater weight to source-text content than to supposed target-language patterns, as predicted by Snell-Hornby.

Section 5.2.4 also provides a certain amount of evidence to confirm Kloepfer's famous hypothesis about the link between boldness and ease of translation, thus tentatively concluding that bold metaphorical expressions may well be translated with less difficulty than other more conventionalised ones, within this type of text and/or subject area at any rate.

7.1.3 Classifying Translation Procedures

Most lists of translation procedures (such as Newmark's) have been practical in nature, emphasising the modifications that translators have consciously introduced, but a more recent tendency was also identified to try and base such a list on more theoretical concepts from metaphor research. This section offers a contribution to this latter type in the hope of being able to go beyond standard classifications of translation procedures by seeking isomorphisms between metaphor parameters and observed translation behaviours. In this way, the list that will be presented next will combine procedures that are specific to metaphor in translation with a few of the more general ones that have been identified in the course of the analysis. Indeed, the parameters have enabled us to examine the same data from six different viewpoints and have therefore brought to light many different general insights and specific observations about what happens to metaphorical language in translation.

While one or two of the procedures are general in nature, most are specific to a particular parameter. The first three items relate to mappings, and the remaining eight focus on individual metaphorical expressions. The list brings together the most frequently observed shifts within different parameters and is drawn from the parameters of mapping, provenance and richness, but with no items derived from existence category, purpose or conventionality. Other procedures that have been recorded but are not considered to be of sufficient prominence have not been included, nor have those that are held to be tantamount to retention (e.g. translation of an image schema plus rich image by an expression that involves the same image schema and rich image), unless it is thought that they reveal something fundamental about metaphor in translation (such as number five). As discussed in Section 3.2.2.5, the term 'procedure' is preferred here, even though many of the items included do not possess the clear intentionality that might normally be expected given the use of this term.

The new list of procedures that emerge from the analysis of the data is as follows:

1. Scattering rather than shifting of mappings that are relatively significant in the source text so that the extent of their use is reduced in translation
2. Modification of a particular expression to create a new one-off mapping (i.e. change of mapping by use of a new source domain)
3. Accentuation of a marginal mapping
4. More generally, modification of a particular expression to cause it to be shifted to another mapping
5. Retention of the source-text shape of a metaphorical expression rather than its replacement by an equivalent deemed to be more natural in the target language
6. Explicitation of a metaphorical expression (including the addition of signalling)
7. Implicitation of a metaphorical expression (including the removal of signalling)
8. Addition of a metaphorical expression, either by converting a non-metaphor to a metaphor or by inserting a metaphor where there was nothing in the original
9. Translation of an image schema plus rich image by an expression that involves the same image schema plus a different rich image (or propositional knowledge)
10. Replacement of a vivid propositional knowledge-based expression with one worded in a more neutral manner, thus giving rise to a more generalised knowledge structure; in many cases, this entails the removal of the metaphorical element
11. Replacement of a rich image by a non-rich image, entailing loss of detail and/or specificity and frequently occurring in conjunction with a change of mapping

It is worth pointing out that I found no clear evidence of any interconnections between the different parameters, which do appear to be independent of each other to all intents and purposes. It is important that this list should be considered a starting point rather than a definitive, finished product. Furthermore, it does not in its present form claim to account equally for all the examples in the corpus: some are not explained at all and others are possibly covered by more than one procedure.

7.2 Recommendations for Further Research

Given the complexity of the project, there are many possible ways in which this line of research could be taken further, with the following being a selection of specific suggestions:

1. Focus on a particular thematic area, such as genetic material or the modelling of machines as people.

2. Move on to other parameters (e.g. basic vs. sub- or superordinate level, conceptual blends, or indeed others—such as explicitation, simplification or culture-embeddedness—that are derived from translation studies).
3. Identify a smaller number of examples that appear to have been problematic for translators and look at them in detail.
4. Investigate other scientific genres, such as textbooks and research papers, in addition to popular scientific texts.
5. Take a single common mapping and trace how it is reflected in target texts.
6. Carry out a close textual analysis of specific extracts to examine metaphorical texture.
7. Produce bilingual studies that would be able to focus on editorial policy, translation as manipulation of meaning or the influence of the cultural setting of a particular language community on translation.
8. Focus on specific text-based metaphor phenomena such as metaphorical terminology, complex metaphor, mixed metaphor, allusion-based metaphor, punning-based metaphor, visual metaphor, extended metaphor and/or chains of metaphorical expressions.
9. Use keywords (such as *circuitry*, *code*, *map*, *population* and *tool*, or the humanisations *try*, *attempt*, *perceive*, *ignore*, *choose*, *recognise* and *know*, within the context of my research) as a way of investigating metaphorical expressions.

The list only represents a few possibilities within what is a very fertile area and could potentially be considerably extended. Since the project has been focused mainly on translation studies, this section restricts itself to suggestions for that discipline only.

7.3 Final Word

Metaphor in translation is a topic that is undeniably complex but represents a vital and growing area of translation studies research. The subject is beginning to yield up its secrets, but in spite of a new and more concerted research effort, there is a great deal that still remains for us to discover. Every project has a particular context and specific emphases, but the main thing is that each one should contribute to our understanding of metaphor in translation and hence, ultimately, of translation in general. Nevertheless, and in spite of the tangible conclusions that the book has reached, there is much that occurs to metaphor in translation that is not yet possible to explain. In line with this, it is just possible that the book has raised more questions than it has answered.

Bibliography

a) Source and Target Texts

The original English versions of each article are listed first, followed by listings in each of the target languages. Names of translators are given where known. I have decided to designate both source-language and target-language versions just by year of publication (rather than differentiating between them as 2004a, 2004b, etc.) as the context will always make it clear which language version is being referred to.

English

Fields, R. Douglas (2004) 'The Other Half of the Brain', in *Scientific American* 290:4 (April), 54–61.

Freeland, Stephen J. & Laurence D. Hurst (2004) 'Evolution Encoded', in *Scientific American* 290:4 (April), 84–91.

Gibbs, W. Wayt (2004) 'Synthetic Life', in *Scientific American* 290:5 (May), 74–81.

Javitt, Daniel C. & Joseph T. Coyle (2004) 'Decoding Schizophrenia', in *Scientific American* 290:1 (January), 48–55.

Nestler, Eric J. & Robert C. Malenka (2004) 'The Addicted Brain', in *Scientific American* 290:3 (March), 78–85.

Solms, Mark (2004) 'Freud Returns', in *Scientific American* 290:5 (May), 28–34.

Traditional Chinese

Fields, R. Douglas (2004) '魅影腦細胞' ['Phantom Brain Cells'], tr. Rongqi Huang, in *科學人* 27 (May), 35–43.

Freeland, Stephen J. & Laurence D. Hurst (2004) '遺傳密碼的演化' ['The Evolution of the Heredity Code'], tr. Qianyi Li, in *科學人* 27 (May), 76–84.

Gibbs, W. Wayt (2004) '組裝生命' ['Assembled Life'], tr. Qianyi Li, in *科學人* 28 (June), 87–97.

Javitt, Daniel C. & Joseph T. Coyle (2004) '透視精神分裂症' ['Penetrating Schizophrenia'], tr. Zhenze Pan, in *科學人* 24 (February), 48–56.

Nestler, Eric J. & Robert C. Malenka (2004) '成癮的大腦' ['The Addicted Brain'], tr. Rongqi Huang, in *科學人* 26 (April), 44–52.

Solms, Mark (2004) '佛洛伊德重出江湖' ['The Reappearance of Freud'], tr. Rongqi Huang, in *科學人* 28 (June), 46–54.

French

Fields, R. Douglas (2004) 'La moitié oubliée du cerveau' ['The Forgotten Half of the Brain'], in *Pour la Science* 323 (September), 56–61.
Freeland, Stephen J. & Laurence D. Hurst (2004) 'Le code de l'évolution' ['The Evolution Code'] *Pour la Science* 319 (May), 60–5.
Javitt, Daniel C. & Joseph T. Coyle (2004) 'Décrypter la schizophrénie' ['Deciphering Schizophrenia'], in *Pour la Science* 316 (February), 56–63.
Nestler, Eric J. & Robert C. Malenka (2004) 'Les drogues et le cerveau' ['Drugs and the Brain'], in *Pour la Science* 318 (April), 42–8.
Solms, Mark (2004) 'Psychanalyse et neurosciences' ['Psychoanalysis and Neurosciences'], in *Pour la Science* 324 (October), 76–81.

German

Fields, R. Douglas (2004) 'Die unbekannte Seite des Gehirns' ['The Unknown Side of the Brain'], in *Spektrum der Wissenschaft* September, 46–56.
Freeland, Stephen J. & Laurence D. Hurst (2004) 'Der raffinierte Code des Lebens' ['The Refined Code of Life'], *Spektrum der Wissenschaft* July, 86–93.
Gibbs, W. Wayt (2004) 'Künstliche Biomaschinen' ['Artificial Biomachines'], in *Spektrum der Wissenschaft* October, 68–75.
Javitt, Daniel C. & Joseph T. Coyle (2004) 'Wenn Hirnsignale verrückt spielen' ['When Brain Signals Go Haywire'], in *Spektrum der Wissenschaft* December, 62–8.
Nestler, Eric J. & Robert C. Malenka (2004) 'Das süchtige Gehirn' ['The Addicted Brain'], in *Spektrum der Wissenschaft* June, 34–41.
Solms, Mark (2004) 'Freuds Wiederkehr' ['Freud's Return'], in *Spektrum der Wissenschaft* October, 76–83.

Italian

Fields, R. Douglas (2004) 'L'altra metà del cervello' ['The Other Half of the Brain'], tr. Cristina Serra, in *Le Scienze* 430 (June), 48–56.
Freeland, Stephen J. & Laurence D. Hurst (2004) 'Il codice dell'evoluzione', tr. Massimo Scaglione, in *Le Scienze* 429 (May), 88–95.
Gibbs, W. Wayt (2004) 'Vita sintetica' ['Synthetic Life'], tr. Massimo Scaglione, in *Le Scienze* 430 (June), 76–84.
Javitt, Daniel C. & Joseph T. Coyle (2004) 'Capire la schizofrenia' ['Understanding Schizophrenia'], tr. Cristina Serra, in *Le Scienze* 426 (February), 58–65.

Polish

Fields, R. Douglas (2004) 'Niedoceniane komórki mózgu' ['The Neglected Brain Cells'], in *Świat nauki* May, 26–34.
Freeland, Stephen J. & Laurence D. Hurst (2004) 'Ewolucja języka genów' ['The Evolution of the Language of Genes'], in *Świat nauki* May, 56–63.
Gibbs, W. Wayt (2004) 'Sztuczne życie' ['Synthetic Life'], in *Świat nauki* June, 58–65.

Javitt, Daniel C. & Joseph T. Coyle (2004) 'Schizofrenia rozkodowana' ['Schizophrenia Decoded'], in *Świat nauki* July, 34–41.
Nestler, Eric J. & Robert C. Malenka (2004) 'Zniewolony mózg' ['The Brain in Bondage'], in *Świat nauki* April, 54–61.
Solms, Mark (2004) 'Freud wiecznie żywy' ['Freud Lives Eternally'], in *Świat nauki* June, 74–81.

Russian

Fields, R. Douglas (2004) 'Другая часть мозга' ['The Other Part of the Brain'], in *В мире науки* July, 22–31.
Freeland, Stephen J. & Laurence D. Hurst (2004) 'Закодированная эволюция' ['Encoded Evolution'], in *В мире науки* July, 54–61.
Gibbs, W. Wayt (2004) 'Синтетическая жизнь' ['Synthetic Life'], in *В мире науки* August, 46–53.
Javitt, Daniel C. & Joseph T. Coyle (2004) 'В глубь расщепленного сознания' ['Into the Depths of the Split Consciousness'], in *В мире науки* April, 30–37.
Nestler, Eric J. & Robert C. Malenka (2004) 'Мозг наркомана' ['The Brain of an Addict'] *В мире науки* June, 50–7.
Solms, Mark (2004) 'Фрейд возвращается' ['Freud Returns'], in *В мире науки* August 2004, 54–61.

b) Other Sources

Abu Ain, Sana Ahmad Abdul-Aziz (2014) *Equivalence in Translating Metaphors and Idioms in King Hussein's Political Speeches*, Unpublished PhD thesis, University of Leeds.
Aleksander, Igor (2000) *How to Build a Mind*, London: Weidenfeld & Nicolson.
Alexieva, Bistra (1997) 'There Must Be Some System in This Madness. Metaphor, Polysemy and Wordplay in a Cognitive Linguistics Framework', in Dirk Delabastita (ed) *Traductio: Essays on Punning and Translation*, Manchester: St. Jerome Publishing, 137–54.
Al-Harrasi, Abdulla (2001) *Metaphor in (Arabic-into-English) Translation with Specific Reference to Metaphorical Concepts and Expressions in Political Discourse*, Unpublished PhD thesis, Aston University.
Al-Hasnawi, A. (2007) 'A Cognitive Approach to Translating Metaphors', in *Translation Journal* 11:3, available at http://translationjournal.net/journal/41metaphor.htm (accessed 17 October 2016).
Alliance for Audited Media (2016) *AAM: Total Circ for Consumer Magazines: Circulation Averages for the Six Months Ended 6/30/2016*, available at http://abcas3.auditedmedia.com/ecirc/magtitlesearch.asp (accessed 17 October 2016).
Alshunnag, M. B. M. (2016) *Translating Conceptual Metaphor in Popular Biomedical Texts from English to Arabic*, Unpublished PhD thesis, University of Salford.
Alvarez, Antonia (1993) 'On Translating Metaphor', in *Meta* 38:3, 479–90.
Arbib, Michael A. (2012) *How the Brain Got Language: The Mirror System Hypothesis*, Oxford: Oxford University Press.
Baker, Mona (2006) *Translation and Conflict: A Narrative Account*, London: Routledge.

Barfield, Owen (1973) *Poetic Diction: A Study in Meaning*, Middletown, CT: Wesleyan University Press (first published 1928).
Becher, V. (2011) *Explicitation and Implicitation in Translation. A Corpus-based Study of English-German and German-English Translations of Business Texts*, Unpublished PhD thesis, University of Hamburg.
Beekman, John & John Callow (1974) *Translating the Word of God*, Grand Rapids, MI: Zondervan.
Bennett, Karen (2011) 'The Scientific Revolution and Its Repercussions on the Translation of Technical Discourse', in Myriam Salama-Carr & Maeve Olohan (eds) *Science in Translation,* special issue of *The Translator* 17:2, 189–210.
Bennett, Karen (2013) 'English as a Lingua Franca in Academia: Combating Epistemicide through Translator Training', in *The Interpreter and Translator Trainer* 7:2, 169–93.
Berber Sardinha, T. (2006) 'A Tagger for Metaphors', Paper given at the Sixth Researching and Applying Metaphor (RAAM) Conference, Leeds University.
Bernardini, Silvia, Marco Baroni & Stefan Evert (2006) 'A WaCky Introduction', in Marco Baroni & Silvia Bernardini (eds) *WaCky! Working Papers on the Web as Corpus*, Bologna: Gedit Edizioni, available at http://wackybook.sslmit.unibo.it (accessed 17 October 2016), 9–40.
Black, Max (1962) *Models and Metaphors: Studies in Language and Philosophy*, Ithaca, NY: Cornell University Press.
Bowker, Lynne & Jennifer Pearson (2002) *Working with Specialized Language: A Practical Guide to Using Corpora*, London & New York: Routledge.
Boyd, Richard (1993) 'Metaphor and Theory Change: What Is "Metaphor" a Metaphor for?', in Andrew Ortony (ed), 481–532.
Braithwaite, Jason (2006) *Seven Fallacies of Thought and Reason: Common Errors in Reasoning and Argument from Pseudoscience*, available at http://www.assap.ac.uk/newsite/Docs/old%20Seven%20fallaciesv5.pdf (accessed 17 October 2016).
Brevik, N. (2008) *Translation Theory with Regards to Translating Metaphors*, available at http://www.proz.com/doc/1831 (accessed 26 October 2016).
Brooke-Rose, Christine (1958) *A Grammar of Metaphor*, London: Mercury Books.
Caballero, Rosario (2002) 'Understanding Is Seeing: A Discourse Perspective on Image Metaphor', in *Cuadernos de Investigación Filológica* 27–28:2001–2002, 81–102.
Caballero, Rosario (2003a) 'Talking about Space: Image Metaphor in Architectural Discourse', in *Annual Review of Cognitive Linguistics* 1, 87–105.
Caballero, Rosario (2003b) 'Metaphor and Genre: The Presence and Role of Metaphor in the Building Review', in *Applied Linguistics* 24:2, 145–67.
Caballero, Rosario (2006) *Re-Viewing Space: Figurative Language in Architects' Assessment of Built Space*, Berlin & New York: Mouton de Gruyter (Applications of Cognitive Linguistics 2).
Caballero, Rosario & Javier E. Díaz-Vera (eds) (2013) *Special Issue: Metaphor and Culture: A Relationship at a Crossroads?, International Pragmatics* 10:2.
Cameron, Lynne (1999) 'Identifying and Describing Metaphor in Spoken Discourse', in Lynne Cameron & Graham Low (eds), 105–32.
Cameron, Lynne & Graham Low (eds) (1999) *Researching and Applying Metaphor*, Cambridge: Cambridge University Press.

Casselman, Bill (1998) *A Dictionary of Medical Derivations: The Real Meaning of Medical Terms*, New York & London: The Parthenon Publishing Group.

Catford, J. C. (1965) *A Linguistic Theory of Translation*, London: Oxford University Press.

Chan, Mei Ling (2011) *How Can Idioms be Translated? An Analysis of English to Chinese Translation in 'Scientific American'*, Unpublished MSc dissertation, Imperial College London.

Charteris-Black, Jonathan (2003) 'Speaking with Forked Tongue: A Comparative Study of Metaphor and Metonymy in English and Malay Phraseology', in *Metaphor and Symbol* 18:4, 289–310.

Charteris-Black, Jonathan (2004) *Corpus Approaches to Critical Metaphor Analysis*, Basingstoke: Palgrave MacMillan.

Charteris-Black, Jonathan (2005) *Politicians and Rhetoric: The Persuasive Power of Metaphor*, Basingstoke: Palgrave Macmillan.

Charteris-Black, Jonathan (2013) *Analysing Political Speeches: Rhetoric, Discourse and Metaphor*, Basingstoke: Palgrave MacMillan.

Crisp, Peter, John Heywood & Gerard J. Steen (2002) 'Metaphor Identification and Analysis, Classification and Quantification', in *Language and Literature* 11:1, 55–69.

Croft, William & D. Alan Cruse (2004) *Cognitive Linguistics*, Cambridge: Cambridge University Press (Cambridge Textbooks in Linguistics).

Crofts, Marge (1982) 'Classification of Metaphors', in *Notes on Translation* 90, 16–17.

Crofts, Marge (1988) 'Translating Metaphors', in *Australian Review of Applied Linguistics* 11:1, 47–53.

Dagut, Menachem (1976) 'Can "Metaphor" be Translated?', in *Babel* 22:1, 21–33.

Dagut, Menachem (1978) *Hebrew-English Translation: A Linguistic Analysis of Some Semantic Problems*, Haifa: University of Haifa.

Dagut, Menachem (1987) 'More about the Translatability of Metaphor', in *Babel* 33:2, 77–83.

Dancygier, Barbara & Eve Sweetser (2014) *Figurative Language*, Cambridge: Cambridge University Press.

Darwin, Charles (1872) *The Origin of Species by Means of Natural Selection, or the Preservation of Favoured Races in the Struggle for Life* (6th Edition), London: John Murray (first published 1859).

Deane-Cox, Sharon (2014) *Retranslation: Translation, Literature and Reinterpretation*, London: Bloomsbury.

Deignan, Alice (2005) *Metaphor and Corpus Linguistics*, Amsterdam & Philadelphia: John Benjamins Publishing Company (Converging Evidence in Language and Communication Research 6).

Deignan, Alice (2008) 'Corpus Linguistics and Metaphor', in Gibbs, Raymond W. (ed) *The Cambridge Handbook of Metaphor and Thought*, New York: Cambridge University Press, 280–94.

Deignan, Alice, Danuta Gabrys & Agnieszka Solska (1997) 'Teaching English Metaphors Using Cross-Linguistic Awareness Raising Techniques', in *ELT Journal* 51, 353–60.

Deignan, Alice & Liz Potter (2004) 'A Corpus Study of Metaphors and Metonyms in English and Italian', in *Journal of Pragmatics* 36:7, 1231–52.

Deignan, Alice & Elena Semino (2010) 'Corpus Techniques for Metaphor Analysis', in Cameron and Maslen (eds) *Metaphor Analysis: Research Practice in Applied Linguistics, Social Sciences and the Humanities*, London: Equinox, 161–179.

Dickins, James (2005) 'Two Models for Metaphor Translation', in *Target* 17:2, 227–73.

Ding, Yan, Dirk Noël & Hans-Georg Wolf (2010) 'Patterns in Metaphor Translation: A Corpus-Based Case Study of the Translation of FEAR Metaphors between English and Chinese', in Richard Xiao (ed) *Using Corpora in Contrastive and Translation Studies*, Newcastle upon Tyne: Cambridge Scholars Publishing, 40–61.

Divasson, L. & I. León (2006) 'Metaphors in English, French and Spanish Medical Written Discourse', in *Elsevier Encyclopedia of Language and Linguistics* (2nd Edition, Volume 8), Kidlington, Oxford: Elsevier, 58–63.

Dobrovol'skij, Dmitrij & Elisabeth Piirainen (2005) *Figurative Language: Cross-Cultural and Cross-Linguistic Perspectives*, Oxford: Elsevier.

Dodge, Ellen & George Lakoff (2005) 'Images Schemas: From Linguistic Analysis to Neural Grounding', in Beate Hampe & Joseph E. Grady (eds), 57–91.

Doolittle, W. Ford (25 June 1999) 'Phylogenetic Classification and the Universal Tree', in *Science* 284:5423, 2124–8, available at http://www.sciencemag.org/content/284/5423/2124.full (accessed 17 October 2016).

Dunbar, Robin (1995) *The Trouble with Science*, London: Faber.

Duranti, Alessandro (1997) *Linguistic Anthropology*, Cambridge: Cambridge University Press (Cambridge Textbooks in Linguistics).

Eco, Umberto (1995) *The Search for the Perfect Language*, Oxford: Blackwell.

Elmallah, F. H. M. (2008) *Arabic-English Translational Crossover Viewed from a Linguistic/Cultural Perspective with Special Reference to the Major Principles Involved in Translating the Metaphorical Language*, Unpublished PhD thesis, University of Edinburgh.

El-Zeiny, Iman (2011) 'Criteria for the Translation and Assessment of Qur'anic Metaphor: A Contrastive Analytic Approach', in *Babel* 57:3, 247–68.

Engstrøm, Anders (2000) 'Review of *Philosophy in the Flesh: The Embodied Mind and its Challenge to Western Thought*, by George Lakoff and Mark Johnson', in *Metaphor and Symbol* 15:4, 267–74.

Erickson, Marianne T. C. (1999) *Toward a Cognitive Poetics of Translation*, Unpublished PhD thesis, Washington University in St. Louis.

Evans, Ruth (1998) 'Metaphor of Translation', in Mona Baker (ed) *Routledge Encyclopedia of Translation Studies* (1st edition), London & New York: Routledge, 149–53.

Evans, Vyvyan & Melanie Green (2006) *Cognitive Linguistics: An Introduction*, Mahwah, NJ & London: Lawrence Erlbaum Associates.

Faber, Pamela & Carlos Márquez Linares (2005) 'A Three-Level Model of Metaphor for Specialised Communication', in Lew N. Zybatow (ed.) *Translationswissenschaft im interdisziplinären Dialog. Innsbrucker Ringvorlesungen zur Translationswissenschaft III* [*Translation Studies as Interdisciplinary Dialogue: Innsbruck Lecture Series on Translation Studies III*], Frankfurt am Main: Peter Lang, 71–93.

Faiq, Said (1998) 'Handling Metaphor in Sensitive Texts: Contributions from Arabic-English Translation', in *Translatio* 17:4, 224–39.

Fantoni, Barry (ed) (1998) *Private Eye, Colemanballs 9*, London: Private Eye.

Farrell, Tim (2004) *Metaphor in Matthew's Gospel: Its Cognitive Structure, Background, Culture Dependence and Translatability*, Unpublished PhD thesis, University of Roehampton.
Fass, D. (1991) 'Met*: A Method for Discriminating Metonymy and Metaphor by Computer', in *Computational Linguistics* 17:1, 49–90.
Fauconnier, Gilles (1994) *Mental Spaces: Aspects of Meaning Construction in Natural Language*, Cambridge: Cambridge University Press.
Fauconnier, Gilles & Mark Turner (2002) *The Way We Think: Conceptual Blending and the Mind's Hidden Complexities*, New York: Basic Books.
Fouces González, Covadonga G. (2007) 'Tras las Huellas de la Metáfora: Una Aproximación a la Traducción de la Metáfora Literaria desde Presupuestos Culturales' ['In the Traces of Metaphor: A Cultural Studies Approach to the Translation of Literary Metaphor'], in *Hermēneus. Revista de Traducción e Interpretación* 9, 39–60.
Fung, Mary M. Y. (1994) *Translating Poetic Metaphor: Explorations of the Processes of Translating*, Unpublished PhD thesis, University of Warwick.
Fung, Mary M. Y. (1995) 'Translation of Metaphor', in Sin-Wai Chan & David E. Pollard (eds) *An Encyclopaedia of Translation: Chinese-English—English-Chinese*, Hong Kong: The Chinese University Press, 658–71.
Gambier, Yves & Luc van Doorslaer (2016) *Translation Studies Bibliography*, Amsterdam & Philadelphia: John Benjamins Publishing Company, available at https://www.benjamins.com/online/tsb/ (accessed 17 October 2016).
Gaudio, Paula, Agnès Ganichaud, Marion Greenway, Beate Knights, Autumn Linn & Adonis Simpson (eds) (2003) *Translation and Metaphor*, Norwich: University of East Anglia (Norwich Papers Volume 11).
Gibbs, Raymond W. (1994) *The Poetics of Mind: Figurative Thought, Language, and Understanding*, New York: Cambridge University Press.
Gibbs, Raymond W. (1999a) 'Researching Metaphor', in Cameron, Lynne & Graham Low (eds) *Researching and Applying Metaphor*, Cambridge: Cambridge University Press, 29–47.
Gibbs, Raymond W. (1999b) 'Taking Metaphor out of our Heads and Putting it in the Cultural World', in Raymond W. Gibbs & G. J. Steen (eds), 145–65.
Gibbs, Raymond W. (ed) (2016) *Mixing Metaphor*, Amsterdam & Philadelphia: John Benjamins Publishing Company.
Gibbs, Raymond W. & Herbert L. Colston (1995) 'The Cognitive Psychological Reality of Image Schemas and their Transformations', in *Cognitive Linguistics* 6:4, 347–78. (Also available in Dirk Geeraerts (ed.) (2006), *Cognitive Linguistics: Basic Readings*, Berlin & New York: Mouton de Gruyter, 239–68.)
Gibbs, Raymond W. & G. J. Steen (eds) (1999) *Metaphor in Cognitive Linguistics*, Amsterdam & Philadelphia: John Benjamins Publishing Company.
Glendinning, Victoria (1995) *Electricity*, London: Hutchinson.
Goatly, Andrew (1997) *The Language of Metaphors*, London & New York: Routledge.
Goatly, Andrew (2007) *Washing the Brain—Metaphor and Hidden Ideology*, Amsterdam & Philadelphia: John Benjamins Publishing Company (Discourse Approaches to Politics, Society and Culture, Volume 23).
Goatly, Andrew & LLE Project (2005) *Metalude: Metaphor at Lingnan University, Department of English*, available at http://www.ln.edu.hk/lle/cwd/project01/web/home.html and associated pages (accessed 17 October 2016).
Goddard, Cliff (2006) *Ethnopragmatics: Understanding Discourse in Cultural Context*, Berlin & New York: Mouton de Gruyter.

Goddard, Cliff & Anna Wierzbicka (2014) *Words and Meanings: Lexical Semantics across Domains, Languages, and Cultures*, New York & London: Oxford University Press.

Grady, Joseph E. (1997) *Foundations of Meaning: Primary Metaphors and Primary Scenes*, Unpublished PhD thesis, Department of Linguistics, University of California, Berkeley.

Grady, Joseph E., Todd Oakley & Seana Coulson (1999) 'Blending and Metaphor', in Raymond W. Gibbs & G. J. Steen (eds), 101–24.

Gries, Stefan Th. & Anatol Stefanowitsch (eds) (2006) *Corpora in Cognitive Linguistics: Corpus-Based Approaches to Syntax and Lexis*, Berlin & New York: Mouton de Gruyter (Trends in Linguistics: Studies and Monographs 172).

Guldin, Rainer (2010) 'Metaphor as a Metaphor for Translation', in James St. André (ed), 161–91.

Guldin, Rainer (2016) *Translation as Metaphor*, Abingdon: Routledge.

Gutt, Ernst-August (1991) *Translation and Relevance*, Oxford: Blackwell.

Halliday, M. A. K. (1985) *An Introduction to Functional Grammar*, London: Edward Arnold.

Hampe, Beate & Joseph E. Grady (eds) (2005) *From Perception to Meaning: Image Schemas in Cognitive Linguistics*, Berlin & New York: Mouton de Gruyter (Cognitive Linguistics Research 29).

Hanks, Patrick (2006) 'Metaphoricity is Gradable', in Anatol Stefanowitsch & Stefan Th. Gries (eds) *Corpus-Based Approaches to Metaphor and Metonymy*, Berlin & New York: Mouton de Gruyter, 17–35 (Trends in Linguistics: Studies and Monographs 171).

Hatim, Basil (1996) *Communication across Cultures: Translation Theory and Contrastive Text Linguistics*, Exeter: University of Exeter Press.

Hatim, Basil & Ian Mason (1997) *The Translator as Communicator*, London & New York: Routledge.

Hellsten, Iina & Brigitte Nerlich (2011) 'Synthetic Biology: Building the Language for a New Science Brick by Metaphorical Brick', in *New Genetics and Society* 30:4, 375–97.

Hermans, Theo (1985) 'Introduction: Translation Studies and a New Paradigm', in Theo Hermans (ed) *The Manipulation of Literature: Studies in Literary Translation*, London: Croom Helm, 7–15.

Hermans, Theo (1991) 'Translational Norms and Correct Translations', in Kitty M. van Leuven-Zwart & Ton Naaijkens (eds) *Translation Studies: The State of the Art. Proceedings of the First James S. Holmes Symposium on Translation Studies*, Amsterdam: Rodopi, 155–69.

Hermans, Theo (1999) *Translation in Systems: Descriptive and System-Oriented Approaches Explained*, Manchester: St. Jerome Publishing.

Hermans, Theo (2007) *The Conference of the Tongues*, Manchester: St. Jerome Publishing.

Hofstadter, Douglas R. (1997) *Le Ton Beau de Marot: In Praise of the Music of Language*, London: Bloomsbury.

Holdcroft, David (1991) *Saussure: Signs, System, and Arbitrariness*, Cambridge: Cambridge University Press.

Holmes, James S. (2004) 'The Name and Nature of Translation Studies', in Lawrence Venuti (ed) *The Translation Studies Reader* (2nd Edition), London: Routledge, 180–92.

Holm, Nanna (2001) *Metaphor and Translation: A Comparative Study of Two Translations into Danish of Federico García Lorca's 'Poeta en Nueva York'*, Unpublished Master's thesis, University of Aarhus.

Hönig, Hans G. (1997a) 'Positions, Power and Practice: Functionalist Approaches and Translation Quality Assessment', in *Current Issues in Language and Society* 4:1, 6–34 (issue entitled *Translation and Quality*, edited by Christina Schaeffner).

Hönig, Hans G. (ed) (1997b) *Konstruktives Übersetzen* [*Constructive Translation*] (2nd Edition), Tübingen: Stauffenburg.

Hoorickx-Raucq, Isabelle (2005) 'Mediating the Scientific Text. A Cultural Approach to the Discourse of Science in some English and French Publications and TV Documentaries', in *The Journal of Specialised Translation* 3, 97–108, available at http://www.jostrans.org/issue03/art_hoorickx_raucq.pdf (accessed 17 October 2016).

Hug, Laura A., Brett J. Baker, Karthik Anantharaman, Christopher T. Brown, Alexander J. Probst, Cindy J. Castelle, Cristina N. Butterfield, Alex W. Hernsdorf, Yuki Amano, Kotaro Ise, Yohey Suzuki, Natasha Dudek, David A. Relman, Kari M. Finstad, Ronald Amundson, Brian C. Thomas & Jillian F. Banfield (2016) 'A New View of the Tree of Life', in *Nature Microbiology* 1, Article number 16048, available at http://www.nature.com/articles/nmicrobiol201648 (accessed 17 October 2016).

Ilisei, Iustina (2012) *A Machine Learning Approach to the Identification of Translational Language: An Inquiry into Translationese Learning Models*, Unpublished PhD thesis, University of Wolverhampton.

Jäkel, Olaf (1999) 'Kant, Blumenberg, Weinrich: Some Forgotten Contributions to the Cognitive Theory of Metaphor', in Raymond W. Gibbs & G. J. Steen (eds), 9–27.

Jay-Rayon, Laurence (2007) 'Traduire les réseaux métaphoriques chez Nuruddin Farah' ['Translating Metaphorical Networks in Nuruddin Farah'], in *Meta* 52:4, 839–58.

Johnson, Mark (1987) *The Body in the Mind: The Bodily Basis of Meaning, Imagination and Reason*, Chicago & London: University of Chicago Press.

Johnson, Mark (2007) *The Meaning of the Body: Aesthetics of Human Understanding*, Chicago & London: University of Chicago Press.

Johnson-Sheehan, Richard D. (1998) 'Metaphor in the Rhetoric of Scientific Discourse', in John T. Battalio (ed) *Essays in the Study of Scientific Discourse: Methods, Practice, and Pedagogy*, Stamford, CT: Ablex Publishing Corporation, 167–79.

Johnston, John (1999) 'Translation as Simulacrum', in Lawrence Venuti (ed) *Rethinking Translation: Discourse, Subjectivity, Ideology*, London: Routledge, 42–56.

Kanellos, Nicolás & Helvetia Martell (2000) *Hispanic Periodicals in the United States—Origins to 1960: A Brief History and Comprehensive Bibliography*, Houston: Arte Público Press.

Katan, David (1999) *Translating Cultures: An Introduction for Translators, Interpreters and Mediators*, Manchester: St. Jerome Publishing.

Keesing, Roger M. (1985) 'Conventional Metaphors and Anthropological Metaphysics: The Problematics of Cultural Translation', in *Journal of Anthropological Research* 41, 201–17.

Kimmel, Michael (2002) *Metaphor, Imagery, and Culture. Spatialized Ontologies, Mental Tools, and Multimedia in the Making*, Unpublished PhD thesis, University of Vienna, available at http://sammelpunkt.philo.at:8080/6/1/zentralpdf.PDF (accessed 17 October 2016).

Kjär, Uwe (1988) *'Der Schrank seufzt': Metaphern im Bereich des Verbs und ihre Übersetzung*, Göteborg: Acta Universitatis Gothoburgensis.

Kloepfer, Rolf (1967) *Die Theorie der literarischen Übersetzung. Romanisch-deutscher Sprachbereich* [*The Theory of Literary Translation. Romance Languages and German*], Munich: W. Fink (Freiburger Schriften zur Romanischen Philologie 12).

Knowles, Murray & Rosamund Moon (2006) *Introducing Metaphor*, London: Routledge.

Knudsen, Susanne (2003) 'Scientific Metaphors Going Public', in *Journal of Pragmatics* 35, 1247–63.

Koller, Werner (1992) *Einführung in die Übersetzungswissenschaft* [*Introduction to the Science of Translation*] (Revised 4th impression), Heidelberg & Wiesbaden: Quelle & Meyer Verlag (first published 1979).

Kövecses, Zoltán (2002) *Metaphor: A Practical Introduction* (1st Edition), Oxford: Oxford University Press.

Kövecses, Zoltán (2005) *Metaphor in Culture: Universality and Variation*, Cambridge: Cambridge University Press.

Kövecses, Zoltán (2006) *Language, Mind, and Culture: A Practical Introduction*, New York & Oxford: Oxford University Press.

Kövecses, Zoltán (2010) *Metaphor: A Practical Introduction* (2nd Edition), Oxford: Oxford University Press.

Kövecses, Zoltán (2014) 'Conceptual Metaphor Theory and the Nature of Difficulties in Metaphor Translation', in Donna R. Miller & Enrico Monti (eds), 25–39.

Kruger, Alet (1993) 'Translating Metaphors in Narrative Fiction', in *Perspectives: Studies in Translatology* 1993:1, 23–30.

Kurth, Ernst-Norbert (1995) *Metaphernübersetzung. Dargestellt an grotesken Metaphern im Frühwerk Charles Dickens in der Wiedergabe deutscher Übersetzungen*, Frankfurt: Lang.

Kyratzis, Sakis (2003) *A New Metaphor for Metaphor: Evidence for a Single Dynamic Metaphorical Category*, Unpublished manuscript.

Lakoff, George (1987a) *Women, Fire and Dangerous Things: What Categories Reveal about the Mind*, Chicago & London: University of Chicago Press.

Lakoff, George (1987b) 'Image Metaphors', in *Metaphor and Symbolic Activity* 2:3, 219–22.

Lakoff, George (1987c) 'Position Paper on Metaphor', delivered at *Theoretical Issues in Natural Language Processing 3 (TINLAP-3), New Mexico State University, 7–9 January 1987*, 194–7, available at http://ucrel.lancs.ac.uk/acl/T/T87/T87–1039.pdf (accessed 17 October 2016).

Lakoff, George (1993) 'The Contemporary Theory of Metaphor', in Andrew Ortony (ed.), 202–51.

Lakoff, George, Jane Espenson & Alan Schwartz (1991) *Second Draft Copy: Master Metaphor List*, available at http://araw.mede.uic.edu/~alansz/metaphor/METAPHORLIST.pdf (accessed 17 October 2016).

Lakoff, George & Mark Johnson (1980) *Metaphors We Live By*, Chicago & London: University of Chicago Press.

Lakoff, George & Mark Johnson (1999) *Philosophy in the Flesh. The Embodied Mind and its Challenge to Western Thought*, New York: Basic Books.

Lakoff, George & Mark Turner (1989) *More than Cool Reason: A Field Guide to Poetic Metaphor*, Chicago & London: University of Chicago Press.

Langacker, Ronald W. (1976) 'Semantic Representations and the Linguistic Relativity Hypothesis', in *Foundations of Language* 14, 307–57.

Langacker, Ronald W. (1987) *Foundations of Cognitive Grammar, Volume 1: Theoretical Prerequisites*, Stanford: Stanford University Press.

Larson, Mildred L. (1984) *Meaning-based Translation: A Guide to Cross-Language Equivalence*, Lanham: University Press of America.

Lau, Nelson C. & David P. Bartel (2003) 'Censors of the Genome', in *Scientific American* 289:2, 34–41.

Lawton, Graham (2009) 'Uprooting Darwin's Tree', in *New Scientist* 201:2692, 34–9.

Leane, Elizabeth (2007) *Reading Popular Physics: Disciplinary Skirmishes and Textual Strategies*, Aldershot: Ashgate.

Lefevere, André (1992) *Translation, Rewriting, and the Manipulation of Literary Fame*, London: Routledge.

Liao, Min-Hsiu (2007) 'The Translation of Interaction in the Genre of Popular Science: The Case of *Scientific American*', in *Proceedings of Corpus Linguistics Conference 2007, University of Birmingham, 27–30 July 2007*, available at http://www.birmingham.ac.uk/documents/college-artslaw/corpus/conference-archives/2007/23Paper.pdf (accessed 26 October 2016).

Liao, Min-Hsiu (2010) 'Translating Science into Chinese: An Interactive Perspective', in *The Journal of Specialised Translation* 13, 44–60, available at http://www.jostrans.org/issue13/art_liao.pdf (accessed 17 October 2016).

Liao, Min-Hsiu (2011) 'Interaction in the Genre of Popular Science', in *The Translator* 17:2, Special Issue: Science in Translation, 349–68.

List of countries by English-speaking population (2016) Available at https://en.wikipedia.org/wiki/List_of_countries_by_English-speaking_population, date updated 15 October 2016 (accessed 17 October 2016).

Liu, Ting (2011) *The Translation of Articles in Popular Science Magazines: A Case Study of* Huanqiu Kexue *(Global Science)—the Chinese Version of* Scientific American, Unpublished MSc dissertation, Imperial College London.

Lobanova, Marina (2002) *György Ligeti: Style, Ideas, Poetics* (trans. Mark Shuttleworth), Berlin: Verlag Ernst Kuhn.

Lovejoy, Arthur Oncken (1936) *The Great Chain of Being: A Study of the History of an Idea*, Cambridge, MA: Harvard University Press.

Low, Graham (1999) 'This Paper Thinks . . . Investigating the Acceptability of the Metaphor AN ESSAY IS A PERSON', in Cameron, Lynne & Graham Low (eds) *Researching and Applying Metaphor*, Cambridge: Cambridge University Press, 221–48.

Low, Graham, Zazie Todd, Alice Deignan & Lynne Cameron (eds) (2010) *Researching and Applying Metaphor in the Real World*, Amsterdam & Philadelphia: John Benjamins Publishing Company.

Maalej, Zouhair (2008) 'Translating Metaphor between Unrelated Cultures: A Cognitive-Pragmatic Perspective', in *Sayyab Translation Journal* 1, 60–82.

Mahon, James Edwin (1999) 'Getting Your Sources Right: What Aristotle *Didn't* Say', in Lynne Cameron & Graham Low (eds), 69–80.

Malmkjær, Kirsten (2005) *Linguistics and the Language of Translation*, Edinburgh: Edinburgh University Press.

Mandelblit, Nili (1995) 'The Cognitive View of Metaphor and its Implications for Translation Theory', in Marcel Thelen & Barbara Lewandowska-Tomaszczyk (eds) *Translation and Meaning PART 3*, Maastricht: Hogeschool Maastricht, 483–95.

Mandelblit, Nili (1997) *Grammatical Blending: Creative and Schematic Aspects in Sentence Processing and Translation*, Unpublished PhD thesis, University of

California, San Diego, available at http://www.cogsci.ucsd.edu/~faucon/NILI/contents.html (accessed 17 October 2016).
Mason, Kirsten (1982) 'Metaphor and Translation', in *Babel* 28:3, 140–9.
Mason, Z. (2004) 'CorMet: A Computational, Corpus-based Conventional Metaphor Extraction System', in *Computational Linguistics* 30:1, 23–44.
Mauranen, Anna (ed.) (2004) *Translation Universals: Do They Exist?*, Amsterdam & Philadelphia: John Benjamins Publishing Company.
Mauranen, Anna (2008) 'Universal Tendencies in Translation', in Gunilla Anderman & Margaret Rogers (eds) *Incorporating Corpora: The Linguist and the Translator*, Clevedon: Multilingual Matters, 32–48.
McVaugh, Michael (2001) 'Cataracts and Hernias: Aspects of Surgical Practice in the Fourteenth Century', in *Medical History* 45:3, 319–40.
Miller, Donna R. & Enrico Monti (eds) (2014) *Tradurre Figure / Translating Figurative Language*, Quaderni del Centro di Studi Linguistico-Culturali, Atti di Convegni CeSLiC—3, Bologna: AMSActa, available at http://amsacta.unibo.it/4030/.
Miller, George A. (1993) 'Images and Models, Similes and Metaphors' in Andrew Ortony (ed.), 357–400.
Mithen, Steven (1996) *The Prehistory of the Mind: A Search for the Origins of Art, Religion and Science*, London: Phoenix.
Mithen, Steven (1998) 'A Creative Explosion? Theory of Mind, Language, and the Disembodied Mind of the Upper Paleolithic', in Steven Mithen (ed) *Creativity in Human Evolution and Prehistory*, London & New York: Routledge, 165–91.
Mithen, Steven (2005) *The Singing Neanderthals: The Origin of Music, Language, Mind and Body*, London: Phoenix.
Montgomery, Scott L. (2013) *Does Science Need a Global Language: English and the Future of Research*, Chicago & London: University of Chicago Press.
Monti, Enrico (2006) 'Dwelling upon Metaphors: The Translation of William Gass's Novellas', in *Nordic Journal of English Studies* 5:1, 117–32.
Mooij, J. J. A. (1976) *A Study of Metaphor. On the Nature of Metaphorical Expressions, with Special Reference to their Reference*, Amsterdam: North-Holland Publishing Co.
Müller, Cornelia (2008) *Metaphors Dead and Alive, Sleeping and Waking: A Dynamic View*, Chicago & London: University of Chicago Press.
Munday, Jeremy (2011) 'Looming Large: A Cross-Linguistic Analysis of Semantic Prosodies in Comparable Reference Corpora', in A. Kruger, K. Wallmach & J. Munday (eds) *Corpus-Based Translation Studies: Research and Applications*, London & New York: Continuum, 169–86.
Munday, Jeremy (2012) *Introducing Translation Studies: Theories and Applications* (3rd Edition), London & New York: Routledge.
Musolff, Andreas, Fiona MacArthur & Giulio Pagani (eds) (2014) *Metaphor and Intercultural Communication*, London: Bloomsbury Academic.
Najjar, Sumaya Ali (2012) *Metaphors in Translation: An Investigation of a Sample of Quran Metaphors with Reference to three English Versions of the Quran*, Unpublished PhD thesis, Liverpool John Moores University.
Nasti, Chiara (2012) *Images of the Lisbon Treaty Debate in the British Press: A Corpus-based Approach to Metaphor Analysis*, Newcastle: Cambridge Scholars.
Newmark, Peter (1977) 'Communicative and Semantic Translation', in *Babel* 23:4, 163–80.
Newmark, Peter (1980) 'The Translation of Metaphor', in *Babel* 26:2, 93–100.

Newmark, Peter (1985) 'The Translation of Metaphor', in W. Paprotté & R. Dirven (eds) *The Ubiquity of Metaphor*, Amsterdam & Philadelphia: John Benjamins Publishing Company, 295–326.

Newmark, Peter (1988a) *Approaches to Translation*, New York: Prentice Hall.

Newmark, Peter (1988b) *A Textbook of Translation*, New York: Prentice Hall.

Nida, Eugene A. (1964) *Toward a Science of Translating: With Special Reference to Principles and Procedures Involved in Bible Translating*, Leiden: E. J. Brill.

Niranjana, Tejaswini (1992) *Siting Translation: History, Post-Structuralism and the Postcolonial Context*, Berkeley: University of California Press.

Norwick, Stephen A. (2006) *The History of Metaphors of Nature: Science and Literature from Homer to Al Gore* (Volumes I–II), Lewiston, Queenston, Lampeter: The Edwin Mellen Press.

Obeidat, Hisham T. B. (1997) *Aspects of the Problems of Translating Metaphor, with Special Reference to Modern Arabic Poetry,* Unpublished PhD thesis, University of St. Andrews.

Olohan, Maeve (2004) *Introducing Corpora in Translation Studies*, London & New York: Routledge.

Olohan, Maeve (2016) *Scientific and Technical Translation*, Abingdon & New York: Routledge.

Omar, Lamis Ismail (2012) *A Cognitive Approach to the Translation of Creative Metaphor in Othello and Macbeth from English into Arabic,* Unpublished PhD thesis, University of Durham.

Ortony, Andrew (ed) (1993a) *Metaphor and Thought* (2nd Edition), Cambridge: Cambridge University Press.

Ortony, Andrew (1993b) 'Metaphor, Language, and Thought', in Andrew Ortony (ed), 1–16.

Pace, Norman R. (2009) 'Mapping the Tree of Life: Progress and Prospects', in *Microbiology and Molecular Biology Reviews* 73:4, 565–76, available at http://mmbr.asm.org/content/73/4/565.full (accessed 17 October 2016).

Papadoudi, Dafni (2010) *Conceptual Metaphor in English Popular Technology and Greek Translation*, Unpublished PhD thesis, University of Manchester.

Pápai, V. (2004) 'Explicitation: A Universal of Translated Texts?' in Anna Mauranen (ed.), 143–64.

Pedersen, Jan (2011) *Subtitling Norms for Television: An Exploration Focussing on Extralinguistic Cultural References*, Amsterdam & Philadelphia: John Benjamins Publishing Company.

Peña Cervel, M. Sandra (2003) *Topology and Cognition: What Image-Schemas Reveal about the Metaphorical Language of Emotions*, München: Lincom Europa.

Pinker, Steven (1994) *The Language Instinct*, New York: William Morrow.

Pisarska, Alicja (1989) *Creativity of Translators: The Translation of Metaphorical Expressions in Non-literary Texts*, Poznań: Uniwersytet im. Adama Mickiewicza w Poznaniu (Seria Filologia Angielska Nr 23).

Pragglejaz Group (2007) 'MIP: A Method for Identifying Metaphorically Used Words in Discourse', in *Metaphor and Symbol* 22:1, 1–39.

Pritchard, Duncan (2010) *What Is this Thing Called Knowledge?* (2nd Edition), London & New York: Routledge.

Pym, Anthony (2000) *Negotiating the Frontier: Translators and Intercultures in Hispanic History*, Manchester: St. Jerome Publishing.

Pym, Anthony (2016) *Translation Solutions for Many Languages: Histories of a Flawed Dream*, London: Bloomsbury Academic.

Reiss, Katharina (1971) *Möglichkeiten und Grenzen der Übersetzungskritik. Kategorien und Kriterien für eine sachgerechte Beurteilung von Übersetzungen*, München: Max Hueber Verlag (Hueber Hochschulreihe 12).

Rennie, John (1996) 'From the Editors: Changing to Stay the Same', in *Scientific American* 274:4, 6.

Rey, Joëlle (2000) 'La traduction des textes scientifiques', in *Target* 12:1, 63–82.

Richards, I. A. (1936) *The Philosophy of Rhetoric*, New York & London: Oxford University Press.

Robinson, Douglas (1997) *Translation and Empire*, Manchester: St. Jerome Publishing.

Rodríguez Márquez, Maria de Montserrat (2010) *Patterns of Translation of Metaphor in Annual Reports in American English and Mexican Spanish*, Unpublished PhD thesis, University of Surrey.

Rohrer, Tim (2005) 'Image Schemata in the Brain', in Beate Hampe & Joseph E. Grady (eds), 165–96.

Romero, E. & Soria, B. (2005) 'The Notion of Grammatical Metaphor in Halliday', in J. L. Martínez-Dueñas Espejo, C. Pérez Basanta, N. Mclaren & L. Quereda Rodríguez-Navarro (eds) *Towards an Understanding of the English Language: Past, Present and Future. Studies in Honour of Fernando Serrano*, Granada: Universidad de Granada, 143–58. [Also available at http://www.ugr.es/~bsoria/papers/Grammatical%20metaphor%20in%20Halliday.PDF (accessed 26 October 2016).]

Rosch, Eleanor (1978) 'Principles of Categorization', in Eleanor Rosch & B. B. Lloyd (eds) *Cognition and Categorization*, Hillsdale, NJ: Lawrence Erlbaum Associates, 27–48.

Sadegh-Zadeh, Kazem (2012) *Handbook of Analytic Philosophy of Medicine*, Dordrecht: Springer.

Sadock, Jerrold M. (1993) 'Figurative Speech and Linguistics', in Andrew Ortony (ed), 42–57.

Sager, Juan C. (1990) *A Practical Course in Terminology Processing*, Amsterdam & Philadelphia: John Benjamins Publishing Company.

Salim-Mohammad, Haris (2007) *Probleme der Metaphernübersetzung aus dem Deutschen ins Arabische am Beispiel des Romans ‚Die Blechtrommel' von Günter Grass [Problems of Metaphor Translation from German to Arabic Based on the Example of the Novel* 'The Tin Drum' *by Günter Grass*], Frankfurt: Peter Lang.

Samaniego Fernández, Eva (1996) *La traducción de la metáfora* [*The Translation of Metaphor*], Valladolid: Servicio de Publicaciones de la Universidad de Valladolid. [Large sections of this work are available at http://pendientedemigracion.ucm.es/info/especulo/numero8/e_saman1.html, http://pendientedemigracion.ucm.es/info/especulo/numero8/e_saman2.html and http://pendientedemigracion.ucm.es/info/especulo/numero9/e_saman3.html (accessed 26 October 2016).]

Samaniego Fernández, Eva (2000) *Diseño y aplicación de un marco de análisis de la traducción de la metáfora* [*Design and Application of a Framework for Analysing the Translation of Metaphor*], Unpublished PhD thesis, Universidad de Alicante, available at http://rua.ua.es/dspace/bitstream/10045/3989/1/Samaniego%20Fern%C3%A1ndez%2c%20Eva.pdf (accessed 26 October 2016).

Samaniego Fernández, Eva (2011) 'Translation Studies and the Cognitive Theory of Metaphor', in *Review of Cognitive Linguistics* 9:1, 262–79.

Santaemilia, José (ed.) (2005) *Gender, Sex and Translation: The Manipulation of Identities*, Manchester: St. Jerome Publishing.

Saussure, Ferdinand de (1986) *Course in General Linguistics* (3rd Edition, translated by Roy Harris), Chicago: Open Court Publishing Company (first published in French in 1916).

Scarpa, Federica (2004) 'Using an Italian Diachronic Corpus for Investigating the "Core" Patterns of the Language of Science', in *Rivista Internazionale di Tecnica della Traduzione* 8, 71–88.

Schaeffner, Christina (1996) *European Integration through Translation?*, Unpublished manuscript.

Schaeffner, Christina (ed) (1999) *Translation and norms*, Clevedon: Multilingual Matters.

Schaeffner, Christina (2004) 'Metaphor and Translation: Some Implications of a Cognitive Approach', in *Journal of Pragmatics* 36:7, 1253–69.

Schaeffner, Christina (2012) 'Finding Space under the Umbrella: The Euro Crisis, Metaphors, and Translation', in *The Journal of Specialised Translation* 17, 250–70, available at http://www.jostrans.org/issue17/art_schaeffner.pdf (accessed 7 January 2017).

Scientific American (2016a) *Kudos and Awards*, available at www.scientificamerican.com/pressroom/about-scientific-american/?tab=kudos-and-awards (accessed 17 October 2016).

Scientific American (2016b) *Quotes*, available at www.scientificamerican.com/pressroom/about-scientific-american/?tab=quotes (accessed 17 October 2016).

Scientific American (2016c) *Key Facts*, available at www.scientificamerican.com/pressroom/about-scientific-american/?tab=key-facts (accessed 17 October 2016).

Scientific American (2016d) *Local Languages*, available at https://www.scientificamerican.com/mediakit/advertise/local-languages/international/ (accessed 17 October 2016).

Semino, Elena (2002) 'A Sturdy Baby or a Derailing Train? Metaphorical Representations of the Euro in British and Italian Newspapers', in *Text* 22, 107–39.

Shabani, A. (2008) *A Comparative Study of the Translation of Image Metaphors of Color in the Shahnameh of Ferdowsi*, available at http://www.translationdirectory.com/articles/article1701.php (accessed 17 October 2016).

Sharkas, Hala (2009) 'Translation Quality Assessment of Popular Science Articles: Corpus Study of the *Scientific American* and Its Arabic Version', in *trans-kom* 2:1, 42–62, available at http://www.trans-kom.eu/bd02nr01/trans-kom_02_01_03_Sharkas_Translation_Quality_Assessment.20090721.pdf (accessed 17 October 2016).

Shuttleworth, Mark (2011) 'Translational Behaviour at the Frontiers of Scientific Knowledge: A Multilingual Investigation into Popular Science Metaphor in Translation', in *The Translator* 17:2, Special Issue: Science in Translation, 301–23.

Shuttleworth, Mark (2013) 'Difficulty in Translation: Grappling with Ligeti's Musical Metaphors', in Helen Julia Minors (ed) *Music, Text and Translation*, London: Bloomsbury Academic, 175–86.

Shuttleworth, Mark (2014) 'Translation Studies and Metaphor Studies: Possible Paths of Interaction between Two Well-established Disciplines', in Donna R. Miller & Enrico Monti (eds), 53–65.

Shuttleworth, Mark & Moira Cowie (1997) *Dictionary of Translation Studies*, Manchester: St. Jerome Publishing.

Simon, Sherry (1996) *Gender in Translation: Cultural Identity and the Politics of Transmission*, London & New York: Routledge.

Snell-Hornby, Mary (1983) *Metaphorical Thought and Translation: Taking a Stand on P. Newmark*, Series A, Paper No. 108, Trier: Linguistic Agency, University of Trier.

Snell-Hornby, Mary (1995) *Translation Studies: An Integrated Approach* (2nd Edition), Amsterdam & Philadelphia: John Benjamins Publishing Company (first published 1988).

Snell-Hornby, Mary (2006) *The Turns of Translation Studies*, Amsterdam & Philadelphia: John Benjamins Publishing Company.

Snell-Hornby, Mary, Franz Pöchhacker & Klaus Kaindl (eds) (1994) *Translation Studies: An Interdiscipline*, Amsterdam & Philadelphia: John Benjamins Publishing Company.

Soares da Silva, Augusto (2013) 'O que sabemos sobre a crise económica, pela metáfora. Conceptualizações metafóricas da crise na imprensa portuguesa' ['What We Know about the Economic Crisis, through Metaphor. Metaphorical Conceptualisations of the Crisis in the Portuguese Press'], in *Revista Media & Jornalismo* 22:1, 11–34.

Stallman, Bob (1999) *Divine Hospitality in the Pentateuch: A Metaphorical Perspective on God as Host*, Unpublished PhD thesis, Westminster Theological Seminary, Philadelphia, available at https://eagle.northwestu.edu/faculty/bob-stallman/dissertation/ (accessed 26 October 2016).

Stambuk, Anuska (1998) 'Metaphor in Scientific Communication', in *Meta* 43:3, 373–9.

St. André, James (ed) (2010) *Thinking Through Translation with Metaphors*, Manchester: St. Jerome Publishing.

Steen, Gerard J. (1999) 'From Linguistic to Conceptual Metaphor in Five Steps', in Raymond W. Gibbs & Gerard J. Steen (eds), 57–77.

Steen, Gerard J. (2014) 'Translating Metaphor: What's the Problem?', in Donna R. Miller & Enrico Monti (eds), 11–24.

Steen, Gerard J., Aletta G. Dorst, J. Berenike Herrmann, Anna Kaal, Tina Krennmayr & Trijntje Pasma (2010) *A Method for Linguistic Metaphor Identification: From MIP to MIPVU*, Amsterdam & Philadelphia: John Benjamins Publishing Company.

Stefanowitsch, Anatol & Stefan Th. Gries (eds) (2006) *Corpus-Based Approaches to Metaphor and Metonymy*, Berlin & New York: Mouton de Gruyter.

Stienstra, Nelly (1993) *YHWH Is the Husband of His People (An Analysis of a Biblical Metaphor with Special Reference to Translation)*, The Netherlands: Kok-Pharos Publishing House.

Talmy, Leonard (1972) *Semantic Structures in English and Atsugewi*, Berkeley: University of California, available at http://linguistics.buffalo.edu/people/faculty/talmy/talmyweb/Dissertation/toc.html (accessed 21 July 2009).

Talmy, Leonard (1975) 'Semantics and Syntax of Motion', in John P. Kimball (ed) *Syntax and Semantics* (Volume 4), New York: Academic Press, 181–238.

Talmy, Leonard (1978) 'Relation of Grammar to Cognition', in Masayoshi Shibatani (ed) *Syntax and Semantics, Volume 6: The Grammar of Causative Constructions*, New York: Academic Press, 43–116.

Talmy, Leonard (1983) 'How Languages Structure Space', in Herbert Pick & Linda Acredolo (eds) *Spatial Orientation: Theory, Research and Application*, New York: Plenum Press, 225–82.

Tcaciuc, Luciana Sabina Mihaela (2013) *Translation Practices at the European Central Bank with Reference to Metaphors*, Unpublished PhD thesis, Aston University.

Tebbit, Simon (2013) *Metaphor in Biblical Translation: A Study of the Translation of Metaphorical Concepts in the Fourth Gospel in Modern Italian Bibles*, Unpublished PhD Thesis, University of Western Australia.

Temmerman, Rita (2002) 'Metaphorical Models and the Translator's Approach to Scientific Texts', in *Linguistica Antverpiensia New Series* 1, 211–26.

Temmerman, Rita (2010) 'Why Special Language Translators Need Insight into the Mechanisms of Metaphorical Models and Figurative Denominations', in Barbara Lewandowska-Tomaszczyk & Marcel Thelen (eds) *Meaning in Translation*, Frankfurt am Main: Peter Lang, pp. 347–65.

Tercedor Sánchez, Maribel, Clara Inés López Rodríguez, Carlos Márquez Linares & Pamela Faber (2012) 'Metaphor and Metonymy in Specialized Language', in Pamela Faber (ed) *A Cognitive Linguistics View of Terminology and Specialized Language*, Berlin and New York: de Gruyter Mouton, 34–72.

Tobias, Shani (2009) 'Untangling and Re-spinning the Web: Translations of Metaphor in Tanizaki's "Shisei"', in *Translation and Interpreting. The International Journal of Translation and Interpreting Research* 1:1, 44–54.

Toury, Gideon (1995) *Descriptive Translation Studies and Beyond*, Amsterdam & Philadelphia: John Benjamins Publishing Company.

Turner, Mark (1991) *Reading Minds: The Study of English in the Age of Cognitive Science*, Princeton, NJ & Oxford: Princeton University Press.

Turner, Mark (1997) *The Literary Mind: The Origins of Thought and Language*, Oxford: Oxford University Press.

Tyulenev, Sergey (2010) 'Translation as Smuggling', in James St. André (ed), 241–74.

Ureña Gómez-Moreno, José Manuel (2011) *Metaphor in Specialised Language: An English-Spanish Comparative Study in Marine Biology*, Unpublished PhD thesis, University of Granada, available at http://hera.ugr.es/tesisugr/19655095.pdf (accessed 17 October 2016).

Ureña Gómez-Moreno, José Manuel & Pamela Faber (2011) 'Strategies for the Semi-Automatic Retrieval of Metaphorical Terms', in *Metaphor and Symbol* 26:1, 23–52.

Vandaele, Sylvie (2002) 'Traduction Publicitaire Médico-Pharmaceutique et Métaphores Conceptuelles' ['Medical-Pharmaceutical Advertising Translation and Conceptual Metaphors'], in Maia, Belinda, Johann Haller & Margherita Ulrych (eds) *Training the Language Services Provider for the New Millennium*, Porto: Faculdade de Letras da Universidade do Porto, 329–39.

Vandaele, Sylvie & Marie-Claude Béland (2012) 'Les Modes de Conceptualisation des Unités d'Hérédité au XIXe Siècle: Spencer, Haeckel et Elsberg' ['The Modes of Conceptualisation of Units of Heredity in the Nineteenth Century: Spencer, Haeckel and Elsberg'], in *Linguistica Antverpiensia New Series* 11, 227–46.

van den Broeck, Raymond (1981) 'The Limits of Translatability Exemplified by Metaphor Translation', in *Poetics Today* 2:4, 73–87.

van Doorslaer, Luc (2007) 'Risking Conceptual Maps: Mapping as a Keywords-related Tool Underlying the Online *Translation Studies Bibliography*', in *Target* 19:2, 217–33.

van Leuven-Zwart, Kitty M. (1989) 'Translation and Original: Similarities and Dissimilarities, I', in *Target* 1:2, 151–81.

van Vaerenbergh, Leona (1996) 'Fachsprache und metaphorische Sprache im Bereich Bank und Börse. Textanalyse und Übersetzungsprobleme', in *Linguistica Antverpiensia* 29–30, 155–69.

Venuti, Lawrence (2008) *The Translator's Invisibility: A History of Translation* (2nd Edition), Abingdon: Routledge.

Vieira, Else Ribeiro Pires (1999) 'Liberating Calibans: Readings of *Antropofagia* and Haroldo de Campos' Poetics of Transcreation', in Susan Bassnett & Harish Trivedi (eds) *Post-colonial Translation: Theory and Practice*, London: Routledge, 95–113.

Vinay, Jean-Paul & Jean Darbelnet (1958) *Stylistique Comparée du Français et de l'Anglais: Méthode de Traduction*, Paris: Didier.

von Flotow, Luise (1997) *Translation and Gender: Translating in the 'Era of Feminism'*, Manchester: St. Jerome Publishing.

von Flotow, Luise (ed) (2010) *Women and Translation*, Ottawa: University of Ottawa Press.

Waldrop, M. Mitchell (1994) *Complexity*, Harmondsworth: Penguin Books.

Whorf, Benjamin Lee (1956) *Language, Thought, and Reality: Selected Writings of Benjamin Lee Whorf* (edited by John B. Carroll), Cambridge, MA: MIT Press.

Wierzbicka, Anna (1992) *Semantics, Culture, and Cognition: Universal Human Concepts in Culture-Specific Configurations*, New York: Oxford University Press.

Wierzbicka, Anna (1997) *Understanding Cultures through Their Key Words: English, Russian, Polish, German, and Japanese*, New York: Oxford University Press.

Yahia, Mohammed (2012) 'Nature Arabic Edition Launches', in *Nature Middle East*, October 18, 2012, available at http://www.natureasia.com/en/nmiddleeast/article/10.1038/nmiddleeast.2012.149 (accessed 17 October 2016).

Yu, Ning (1998) *The Contemporary Theory of Metaphor: A Perspective from Chinese*, Amsterdam & Philadelphia: John Benjamins Publishing Company.

Zahid, Abdelhamid (2004) *A Model for Metaphor Translation from English Literature Into Arabic*, Marrakech: Imprimerie Papeterie el Watanya.

Zahri, Maysoon (1990) *Metaphor and Translation*, Unpublished PhD thesis, University of Salford.

Zelinsky-Wibbelt, Cornelia (2003) *Text Transfer: Metonymy and Metaphor, Translation and Expert-Lay Communication*, Berlin & New York: Mouton de Gruyter (Text, Translation, Computational Processing 4).

Zhang, Yingying (2009) *Translation of Metaphors in Literary Discourse – An Analysis of Weicheng and Fortress Besieged*, Unpublished Master's Thesis, University of Queensland.

Index

Boldface page locators refer to the page where a table containing that term is located.

For Product Safety Concerns and Information please contact our EU
representative GPSR@taylorandfrancis.com
Taylor & Francis Verlag GmbH, Kaufingerstraße 24, 80331 München, Germany

www.ingramcontent.com/pod-product-compliance
Ingram Content Group UK Ltd.
Pitfield, Milton Keynes, MK11 3LW, UK
UKHW022118190926
475702UK00013B/1519

* 9 7 8 0 3 6 7 3 6 6 3 2 2 *